Beyond the Third Dimension

BEYOND THE THIRD DIMENSION

Geometry, Computer Graphics, and Higher Dimensions

Thomas F. Banchoff

SCIENTIFIC AMERICAN LIBRARY

A Division of HPHLP
New York

Library of Congress Cataloging-in-Publication Data

Banchoff, Thomas F.
    Beyond the third dimension. / Thomas F. Banchoff.
    p.   cm.
    1. Hyperspace.   2. Dimensions.   3. Computer graphics.

ISBN 0-7167-5025-2
ISBN 0-7167-6015-0 (pbk)

I. Title.
QA691.B26   1990   516.3′6—dc20   90-8522

ISSN 1040-3213

Printed in the United States of America

Scientific American Library
A division of HPHLP
New York

Distributed by W. H. Freeman and Company
41 Madison Avenue, New York, New York   10010
Houndmills, Basingstoke RG21 6XS, England

First printing 1996, HAW

This book is number 33 of a series.

CONTENTS

To my wife, Lynore
and our children, Tom, Ann, and Mary Lynn

PREFACE

For well over a century, people have been fascinated by what it means for objects to exist in different dimensions, higher or lower than our own third dimension. This book treats a number of themes that center on the notion of dimensions, tracing the different ways in which mathematicians and others have met them in their work. Although many different branches of mathematics have used the idea of dimensions to gain new insights, it is primarily the geometers who have delighted in imagining phenomena that take place in a whole range of dimensions. Scientists, philosophers, and artists have also found inspiration by considering different dimensions, and many examples of the influence of this concept appear in the following chapters. In recent years a number of excellent books have detailed the importance of dimensional ideas in physics, philosophy, and modern art, and many of these titles are collected at the end as a set of further readings.

The book at hand represents a forty-year personal fascination with a topic that has always presented something new each time I thought I understood it. At first, dimension was only a mysterious word surprising me in a frame from a Captain Marvel comic. As Billy Batson, boy reporter, tours a futuristic laboratory, an

Einstein-like figure proudly states, "This is where our scientists are studying the seventh, eighth, and ninth dimensions." A thought balloon goes up from Billy Batson (and from me), "I wonder what happened to the fourth, fifth, and sixth dimensions?" Shortly after, a *Strange Adventures* comic book introduced me to the classic theme of a being from a higher dimension intruding on our world in the same way that we could intersect the flat world of the surface of a still pond. As we will see, these seemingly bizarre ideas appear again in the serious study of dimensions. Only later did I begin to appreciate the inspiration for these stories in the nineteenth-century classic *Flatland*, which continues to be the best introduction to the interrelationship between worlds of different dimensions.

By the time I began studying geometry in high school, I had discovered that the theme of dimensions is a thread running through all of mathematics, and into the world beyond. Architectural drawings and maps of the world try to reduce three-dimensional information to flat pages, and I became conscious both of their power and their limitations. Formulas for area and volume and formulas from elementary algebra presented patterns relating geometry in the plane to geometry in space, and again and again these patterns would invite me to consider generalizations to a fourth dimension or higher. As I learned new mathematical subjects, I always tried to see what the various notions would mean in different dimensions, but I often became frustrated at my inability to picture or model these higher-dimensional interpretations. Since the nineteenth century, people had been inventing ways to treat phenomena in higher dimensions, and I could share both their excitement about these ideas and their sense of inadequacy in imagining them.

My own opportunity to make a contribution to the visualization of higher dimensions came when I first encountered computer graphics as a young assistant professor at Brown University twenty-three years ago. The ability to see and manipulate complicated three-dimensional forms on a television screen suggested an ideal way to approach the even more complicated forms arising in higher-dimensional geometry. Much of what appears in this book is a description of how computer graphics extends our abilities to visualize different dimensions in ways that were not contemplated just a generation ago. And as this work in geometry parallels research in other fields, the insights we achieve as we investigate patterns in higher dimensions will be more and more useful to researchers in science and in art. We will see some of these influences in this volume, and even more await us in the future.

Many people have contributed to the development of this book, and I have attempted to thank them in a section of acknowledgments at the end. Particular thanks go to the students who generated the new art for this project on computers at Brown University: Davide Cervone, Nicholas Thompson, Jeff Achter, and Matthew Stone.

No book can appear without cooperation of an author and a publisher, but the staff of the Scientific American Library has been exceptional in the extent of collaboration offered throughout the project. If the writing is clear, that is due in large measure to the extraordinarily thorough efforts of editor Susan Moran, and also to the careful reading of project editor Rita Gold. Diane Maass ably supervised the final stages of proof. Credit belongs to Alice Fernandes-Brown for the design, to John Hatzakis for page layout, and to Travis Amos for photo selection. Susan Stetzer coordinated the production of the book. The idea for the project goes back to Neil Patterson, and the greatest credit for helping bring this book into existence goes to my editor and friend, Jeremiah J. Lyons.

Thomas F. Banchoff
April 1990

This edition features new computer graphics illustrations produced by Davide Cervone and the author at the Geometry Center, University of Minnesota. There are also several corrections in response to suggestions of helpful reviewers, in particular, H. S. M. Coxeter. For making this new edition possible, thanks go to editor Jonathan Cobb.

Thomas F. Banchoff
October 1995

Beyond the Third Dimension

1 INTRODUCING DIMENSIONS

As we watch through the lenses of a microscope, an amoeba goes about the course of its virtually two-dimensional life, confined to the narrow region between the slide and its coverslip. We observe from above as the amoeba moves around, encountering other creatures like itself, capturing food, and avoiding predators. Part of the cell membrane forms a line of defense entirely surrounding the amoeba and protecting its nucleus inside from threats by other creatures on the slide. But the words *inside* and *surrounding* do not mean the same to us in three-dimensional space as they do to the inhabitants of this nearly flat space. No amoeba in this space can ever come into direct contact with the nucleus of another. We, however, can look down from another direction entirely and see the very insides of the organism. Not only is the nucleus exposed to our view, but we can also poke it directly, a strange and disturbing event for the surprised amoeba. From our three-dimensional perspective, we visualize the world of the microscope slide in a totally different way than do its inhabitants.

One hundred and six years ago, a brilliantly conceived book exploited this fundamental idea of interaction between creatures of different dimensions to encourage its readers to break the bonds of limited perspective and open their minds to new ways of perceiving. Its author, Edwin Abbott Abbott, was a clergyman and the

Were we able to look down on a two-dimensional universe, we would be able to see every part of a structure at once in the same way that a bird flying overhead could see the complete pattern of a maze, invisible to the lost wanderer within.

headmaster of a school in Victorian England. As a leader in the movement to provide educational opportunities for young men and women of all social classes, he was often frustrated by prevailing social attitudes and by establishment views in education and religion. Of his fifty books, the one that still speaks clearly to our own day is his little masterpiece *Flatland*, simultaneously a social satire and an introduction to the idea of higher dimensions.

Flatland describes an entire race of beings who are two-dimensional, living on a flat plane, unaware of the existence of anything outside their universe. How they lived and interacted and communicated is a fascinating story, and the narrator, A Square, does an excellent job of interpreting his society and his world to us living in what he calls "Spaceland." His task is prodi-

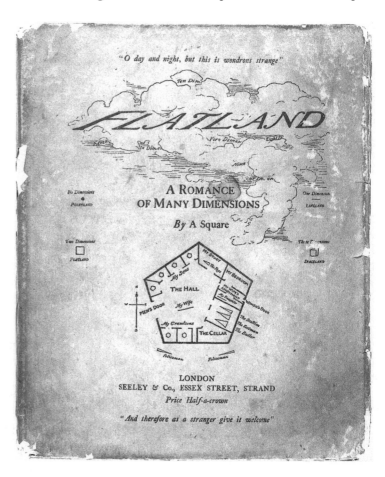

The cover of the 1884 first edition of *Flatland* not only invited the readers into realms of new dimension, but into the two-dimensional house of the book's narrator A Square. Although A Square can see only one room at a time, his house is totally open to our view.

gious, because as difficult as it is for us to imagine how the flat world looks to its citizens, it is truly impossible for the two-dimensional narrator to appreciate the full reality of Spaceland. In particular, he cannot conceive the kind of total view of his existence that we possess. Like a technician watching the movements of the amoeba, we can observe the changing positions of the creatures in Flatland. We can see all parts of a house simultaneously and the contents of any room or any enclosure. From the Flatland point of view, we are omnivident, seeing everything. It is little wonder that A Square, on first hearing about this superior vision, supposes that anyone who possesses it must be divine.

To help A Square understand the all-encompassing view from the third dimension, Abbott presents a dimensional analogy. He asks A Square to imagine what it would be like for him to observe Lineland, a one-dimensional universe populated by segments. A Square would be able to see all creatures in this world at the same time. The King of Lineland, a long segment, would be very surprised if A Square poked his inside without disturbing either of his extremities.

Just as a Flatland creature can view all of Lineland, we in space have a superior view of Flatland. In the story, the power of the analogy makes a great impression on A Square. He asks what it would be like for a being from a fourth dimension to "look down from on high" and see everything in three-space, even the very insides of solid humans. What about worlds of five or six dimensions, each one able to look down on its predecessor and each one open to the all-seeing scrutiny of the next?

Abbott used dimensional analogies to great effect in raising questions about the way we see the world, especially when we come into contact with the truly transcendental. For over a century, mathematicians and others have speculated about the nature of higher dimensions, and in our day the concept of dimension has begun to play a larger and larger role in our conception of a whole range of activities.

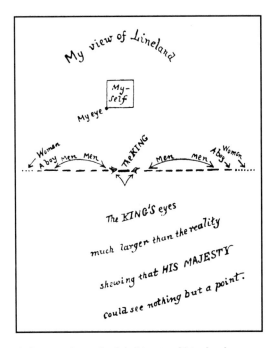

A Square views the inhabitants of Lineland.

The Many Meanings of Dimension

Architects and construction workers for a new mathematics building calculate the amount of carpeting, wiring, and air conditioning necessary to double the size of the entrance area. A team of radiologists examines a sequence of magnetic resonance images displaying a tumor on a patient's optic nerve as it responds to treatment over the course of a month. A group of geologists studying global warming patterns reconstructs the climate history of the Midwest over ten thousand years. A choreographer challenges her students to dance with their backs flat against a wall. In an interactive computer graphics laboratory, a mathematics professor and her student programmers adapt video game technology for use in the study of complex surfaces. As we will see in the chapters to follow, all of these people shape their experiences by exploiting the concept of dimensions.

Although these examples all make use of the notion of dimension, they interpret it differently. The word *dimension* is used in many ways in ordinary speech, and it has several technical meanings as well. When we refer to a "new dimension," it almost always means that we are measuring some phenomenon along a new direction. The word can be used as a metaphor, as for example when we discover that a rather "one-dimensional" colleague is an accomplished guitarist and skeet shooter, giving her two additional "dimensions of personality." In more conventional usage, dimensions are measurements that can specify location—

In a modern computer graphics laboratory, it is possible to investigate the interior structure of a complicated mathematical object. These images represent the so-called pedal surface of a curve in three-dimensional space, formed by the closest points to the origin on all planes that just touch the curve at some point. The red spot on the complete figure is called a "swallowtail catastrophe," and the cutaway views show the structure of the surface near such a point.

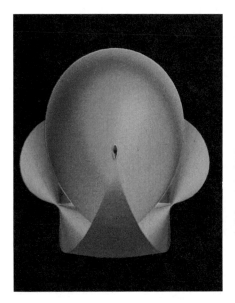
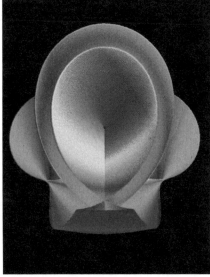
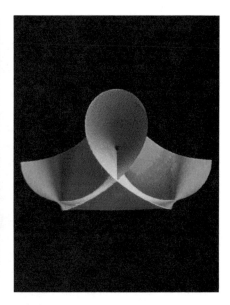

for example the latitude, longitude, and depth of a submarine—or that can specify shape—perhaps the height, top radius, and bottom radius of a tapering flagpole. A list of dimensions may include other kinds of characteristics, as when we specify a brass gong by its weight, thickness, radius, brightness, and tone. We use time and space together as dimensions when we make an appointment to meet someone at nine o'clock on the corner of Fourth Avenue and Twenty-Third Street on the thirty-seventh floor. In recent years, physicists have begun to speak about configurations that involve 11 or 26 dimensions. Mathematicians often speak about structures in n-dimensional space.

One very common way of thinking of dimensions is to look at what engineers call "degrees of freedom." This notion is implicit in much of our day-to-day activity, as in the following scenario. A driver finds herself in a tunnel under Baltimore harbor crawling behind a large truck. "Do Not Change Lanes," she is admonished. She is stuck in one dimension, effectively kept in line, blocked by the vehicle in front of her and the one in back.

Once outside the tunnel, she is again able to move in two dimensions, because she now has an additional "degree of freedom," allowing her to change lanes to the right or to the left. But a bit later she finds all lanes blocked by bridge construction in Havre de Grace. Trucks and cars hem her in on all sides. She wishes she could escape into that inviting third dimension, where a police helicopter hovers unconstrained by the traffic on the roadway. Her degrees of freedom are not limited to spatial dimensions. She may also wish that she had used another kind of dimension to alleviate her problem, the dimension of time. If only she had timed her trip to arrive at the bridge at a slack period with no traffic buildup.

All of these notions of dimension share some characteristics, which we begin to appreciate better as we try to visualize the relationships they represent.

Dimensions as Coordinates

Common to almost all of these ideas of dimensions are lists of numbers, coordinates that specify some quantity associated with an object or a phenomenon. For example, the driver in the Baltimore tunnel can record her position by noting the tenths of a mile from the starting point and the number of feet from the tunnel wall. By far the most familiar example of such coordinates are the length, width, and height of a rectangular box. These three numbers completely specify the shape of a box. Once we know these numbers, we can construct the box, or picture it in our minds even before it is constructed. We can use the familiar concept of the coordinates of a box to help visualize patterns of data in different dimensions—one, two, three, and eventually four and more.

In many applications of mathematics, from the economics of health care to the mapping of distant galaxies, the information we have to process consists of many different measurements for each observation. Making sense out of such complicated collections of data is one of the greatest challenges to social scientists or physical scientists, and it is in this area that the experience of mathematicians concerned with visualizing higher dimensions can be of most help. All observational scientists rely on their ability to recognize trends and patterns, to identify regularities that lead to predictable behavior. Our visual sense is our most powerful faculty for discerning such patterns. One of the most effective ways of bringing our visual faculty into play is to interpret the sequence of measurements for each observation as a point in a space of the appropriate dimension.

One number for each individual is sufficient to record the heights in a family, and all the coordinates can be entered on the same number line, for example the doorjamb at the kitchen entrance. To record both the height and width of the armspan for each family member, we could mark the two quantities on two different number lines, but we get much more information by displaying the data on a two-dimensional surface, a kitchen wall next to the doorjamb. Each family member determines one smallest-fitting rectangle, with the width marking the armspan. The definite advantage of a two-dimensional display is that the single point on the wall indicates both quantities, height and armspan, for each individual. A pair of measurements has led to a two-dimensional quantity. We are better able to see relationships between the quantities when they are displayed on the same two-dimensional diagram.

As an example of such a relationship, for most adults the smallest rectangle that records height and armspan is nearly square. Once we observe this trend, we can reduce the dimension of the system. We do not have to record the armspan since we can deduce it once we know the height. This simple example lies at the heart of the modern subject known as exploratory data analysis.

To record foot size as well as height and armspan for each family member, we determine the smallest rectangular box that contains that person. An upper corner of the box gives a three-dimensional quantity, the record of all three numbers at the same time.

The power of these familiar one-, two-, and three-dimensional frameworks becomes evident when we use them to record observations that have little to do with height or width or length. The three numbers height, weight, and age can be recorded and visualized on the same three-dimensional framework that we used to keep track of the spatial coordinates height, armspan, and foot length.

We have ready-made ways of visualizing data of dimensionality one, two, or three. Marks on a number line, points on a piece of graph paper, points in space are all available to us as a means of picturing a set of coordinates. But what if we were concerned with more than three measurements, say height, armspan, weight, and age? We would have four measurements for each family member, and where would we be able to record them, and how would we visualize the records we create?

Once we see the data laid out on a familiar framework in two or three dimensions, we can identify relationships that are simply invisible if all we can see is a long list of measurements. The coordinate structure forms the backdrop against which we organize our observations and achieve our insights. In order to use visualization techniques effectively in more complicated situations, we need to become familiar with frameworks of even greater dimensionality.

A thread that runs through all considerations of dimensions is the attempt to use insight obtained in one dimension to understand the next. We use this process automatically as we walk around an object or a structure, accumulating sequences of two-dimensional visual images on our retinas from which we infer properties of the three-dimensional objects causing the images. Thinking about different dimensions can make us much more conscious of what it means to see an object, not just as a sequence of images but rather as a form, an ideal object in the mind. We can then begin to turn this imaging faculty to the study of objects that require even more exploration before we can understand them, objects that cannot be built in ordinary space.

Dimensional Progressions

Dimensional analogies did not begin with *Flatland*. The power of comparing insights from different dimensions already occurs explicitly in the works of Plato. In the seventh book of the *Republic*, Socrates converses with Glaucon about the education of the guardians of an ideal state: start with arithmetic and the study of the number line, then proceed to plane geometry, an essential understanding for anyone charged with military defense or the layout of cities. When Socrates asks what should come next, Glaucon suggests astronomy. Socrates chides him for missing an essential step, that of solid geometry, a subject that he considers to be neglected in the schools of his day. Only after proceeding from the first dimension to the second and then to the third would a student be prepared to consider the motions of the heavens.

Plato recognized the built-in progression of dimensions, and he knew how powerful the devices of analogy could be in suggesting solid geometry theorems corresponding to results in plane geometry. But strictly speaking, he did not let the momentum of one to two to three carry him into a spatial fourth dimension. The response to that invitation came only many centuries later, in the early 1800s, when mathematicians in different parts of the world opened their minds to new kinds of geometry. One breakthrough was the discovery of non-Euclidean geometries, satisfying all but one of Euclid's plane geometry axioms. Another crucial insight occurred when mathematicians realized that our plane and solid geometries were just the beginning of a sequence of geometries of higher and higher dimension. Both of these developments challenged the prevailing view that geometry was totally confined to the description of direct physical experience. Inability to visualize what non-Euclidean and higher-dimensional geometries might mean caused many to reject them. Writers like Abbott and Karl Friedrich Gauss and Hermann von Helmholz developed dimensional analogies to enlarge the faculty of imagination so that people could consider the new mathematical creations.

Analogy is surely the dominant idea in the history of the concept of dimensions. If we truly understand a theorem in plane geometry, then we should be able to find one or more analogies in solid geometry, and conversely, solid geometry theorems will often suggest new relationships among plane figures. Theorems about squares should correspond to theorems about cubes or square prisms. Theorems about circles should be analogous to theorems about spheres or cylinders or cones. But if we learn a good deal by going from two dimensions to three, would we not learn even more by going from three dimensions to four?

Mathematicians began to follow different paths within this progression, developing sequences of analogous figures starting even farther back along the dimensional ladder. One possible sequence started with a point, having zero dimensions, no degrees of freedom. A point moving in a straight line generates a segment with two endpoints, a fundamental one-dimensional object. A segment moving perpendicular to itself in a plane generates a figure with four corners, a square, the basic object in the second dimension. Proceeding to the third dimension, we move a square perpendicular to itself to form a cube, a basic three-dimensional object. Even though A Square could no longer appreciate this process fully, he could follow along at a theoretical level and deduce certain properties of this cube that he could not see—for example, that it has eight corners. Next we ask what appears if we move a cube in a fourth direction perpendicular to all its edges. We would get a basic four-dimensional object, a hypercube, and although we can no longer fully appreciate the process, we can predict that such a hypercube will have 16 corners. The numbers of corner points generate a geometric progression, and we can easily arrive at a formula for the number of corners of a cube in any dimension.

A consideration of boundaries leads to another progression. A segment has two boundary points. A square is bounded by four segments. The boundary of a cube is six squares. Following this progression, we expect that the hypercube will be bounded by

The analogue of a cube in any dimension can be generated by moving the preceding lower-dimensional cube perpendicular to itself.

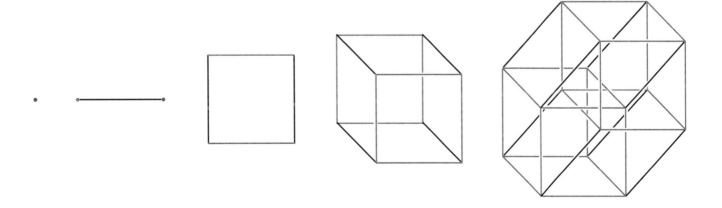

In an 1880 paper, William Stringham used techniques from analytical geometry to produce illustrations showing partially completed, regular four-dimensional figures projected into three-dimensional space.

cubes, and that there will be eight of them. The formula for the number of boundary pieces in any dimension can be found by using an arithmetic progression.

But does a hypercube really exist? Mathematicians did not feel they had to answer that question. They could predict the numbers of corners and boundary pieces of analogues of the cube in any dimension, whether or not they corresponded to some physical reality. However, these lists of numbers still left something to be desired. In plane and solid geometry, the objects were not only real, they could also be represented as diagrams or models, capable of revealing significant relationships. How can we see what a hypercube would look like? If we can't see it, how do we know that our assertions about it are true?

Geometers in the last century devised methods for visualizing objects in higher dimensions, and these methods are worth looking at over the course of this book. Quite sophisticated in some ways but with limitations that were often frustrating, the imaging and modeling techniques of a hundred years ago were inadequate for interpreting complicated objects in four dimensions. Eventually higher-dimensional geometry came to be based not only on analogy but on coordinate geometry, which could translate geometric concepts into numerical and algebraic form. Although such formal methods put the mathematics on a firm footing, they could not satisfy the desire to "see" the higher dimensions.

For the task of visualizing objects and relationships in the fourth and higher dimensions, the ideal instrument is the modern graphics computer.

A Revolution in Visualization Technology

The graphics computer is but the latest in a series of inventions that have enabled us to see in previously inaccessible directions. Four hundred years ago Galileo Galilei turned his newly constructed telescope toward Jupiter and saw moons, an unthinkable vision in a society convinced that all heavenly bodies revolved around the Earth. Now the descendants of Galileo's little optical device reveal the presence of quasars billions of light years away.

A century after Galileo, Anton van Leeuwenhoek's microscope allowed him to explore undreamed-of worlds of very small animals and plants literally within our own blood and tears. Today's powerful electron microscopes show us objects smaller by orders of magnitude, including the structure of genetic material itself.

Early in this century, Wilhelm Conrad Röntgen's discovery of X rays revealed the solid skeleton inside our bodies and gave evidence of the state of the organs functioning there. How much more powerful are today's CAT scans and magnetic resonance imaging, which literally expose to our view slices of our bodies.

These dramatic excursions into the almost unimaginably far away, the equally unfathomably small, and our hidden insides are evidence of our ability to go beyond ourselves, to see the previously unseeable. Equally dramatic is a present-day revolution in our ability to visualize phenomena in other dimensions.

Thanks to striking developments in computer graphics, it is now possible for us to have direct visual experience of objects that exist only in higher dimensions. As we watch images moving on the screen of a graphics computer, we are faced with challenges like those of the first scientists to make use of telescopes or microscopes or X rays. We are seeing things now that have never been visible before, and we are just learning how to interpret these images. It is literally true that we are in the first stages of a new era when it comes to visualizing dimensions.

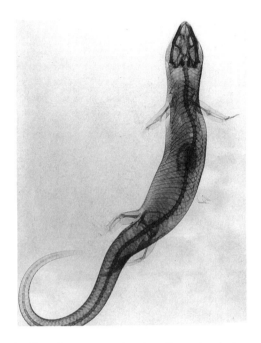

An X-ray image presents a two-dimensional view of the interior of a salamander.

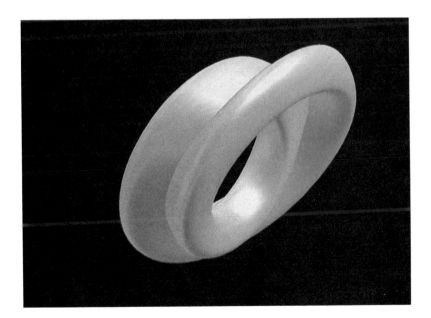

A smoothly shaded and illuminated computer-generated image of a Klein bottle, a twisted surface that cannot be built in ordinary space without intersecting itself, but that can be constructed without self-intersection in four-dimensional space, as described in Chapter 9.

2 SCALING and MEASUREMENT

For many centuries before people considered the concept of higher-dimensional space, they recognized numerical and algebraic patterns in plane and solid geometry. Artisans and scientists and mathematicians developed formulas to describe the regularities that they observed in their measurements, and they knew that the coefficients and exponents that appeared in these formulas were related to the dimensions of the space in which they were working. In a real sense, the concept of dimension became identified with the exponents, with exponent 2 appearing in the formulas of plane geometry and exponent 3 in the formulas of solid geometry. But some of the algebraic patterns that had analogous forms in two- and three-dimensional geometry also had analogues with exponent 4 or higher. What sort of geometry might correspond to these new relationships? The examination of the way numerical and algebraic patterns work in plane and solid geometry paved the way for the appearance of higher-dimensional geometry.

Differences in formulas appear when we consider measurements of analogous objects in different dimensions. We can see the characteristic features of dimensions most clearly when we scale an object up or down. Consider the problem of preparing a photograph for mailing. A square photograph requires a certain

The great pyramids of ancient Egypt remain a source of geometric inspiration. Volume formulas for pyramids contain patterns that continue into higher dimensions.

amount of string and a certain amount of wrapping paper. If we double the size of the square photograph, we need twice as much string and four times as much paper. Doubling the size of a cubical box requires twice as much string, four times as much paper, and eight times as much packing material. Similarly, if we double the size of an entrance hall, then all linear quantities, like the length of wiring, are doubled. But the quantities involving area, like the amount of paint for the walls and the square feet of carpeting for the floor, are multiplied by four, and quantities involving volume, like the cubic feet of space to be handled by the air conditioning units, increase by a factor of eight.

The quantities length, area, and volume express "the amount of material" for objects in different dimensions. The significant fact is that we can determine the dimension simply by looking at the power of two by which the quantity is multiplied when the size is doubled. A quantity like volume is called three-dimensional if it is multiplied by two to the third power when the size of the object is doubled. The two-dimensional quantity, area, is multiplied by two to the second power when the size is doubled, while the one-dimensional quantity, length, is multiplied by just two (or "two to the first power") when the size is doubled. If we encounter a quantity that it is multiplied by 16, two to the fourth power, when the size is doubled, then we would say that the

Enlarging a photograph by doubling all dimensions multiplies the area by four.

quantity is four-dimensional. A five-dimensional quantity would be multiplied by 32 if the size were doubled.

Exponent Patterns for Basic Building Blocks

Patterns that arise in the formulas of plane geometry often correspond to patterns in the formulas of solid geometry, and they may suggest analogous relationships in higher dimensions. One of the simplest of these patterns arises when we measure the length or area or volume of the basic building blocks in any dimension— namely a segment in a line, a square in a plane, and a cube in space. In the line, a segment has length m if we can cover it exactly by m segments of unit length. Similarly in two dimensions, a square with side length m can be filled exactly with m^2 unit squares. And in three dimensions, a cube having side length m can be filled exactly with m^3 unit cubes. The pattern is in the exponents: in dimension n, the volume of an n-cube having sides of length m is m^n, so a four-dimensional cube having sides of length m would be filled with exactly m^4 unit four-dimensional cubes.

But is there a geometric configuration that corresponds to this expression m^4? It would have to be four-dimensional, and indeed it does exist in four-dimensional space, if we interpret the word *exist* in a different way than in ordinary speech. When we speak informally about a square, we think of chalkboard sketches, or

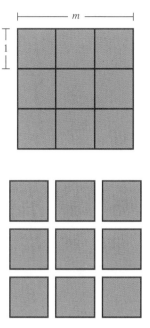

A square of side length m subdivided into m^2 unit squares.

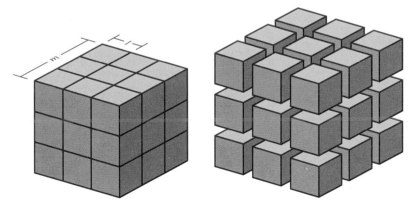

A cube of side length m subdivided into m^3 unit cubes.

much more exact renditions on an architect's table or on a computer-aided drafting device. Yet the formal geometric theorems pertaining to a square are about none of these physical representations, but about the abstract idea of the square, more perfect than anything we can construct. As the followers of Plato might put it, the ideal square exists only in the mind of God. It is such an ideal square that has side length exactly m covered precisely by m^2 unit squares. Similarly the volume formulas refer to perfect cubes, not to the physical representations we see around us. So it is also that the four-dimensional version of this algebraic expression corresponds to an ideal object, a hypercube having side length m, existing in the mind of that same God, and filled with m^4 perfect unit hypercubes. The difference is that we are able to construct in space a model of the m^3 solid cubes, while it is not possible for us to build a similar model of m^4 hypercubes.

Volume Patterns for Pyramids

In nearly all of the important formulas for measuring objects, the different dimensions appear either in the exponents or the coefficients. For volumes of cones and pyramids, the dimensions show up in two different ways.

We do not know which ancient artisan first filled a cone three times from the water in a cylinder and thereby observed that the

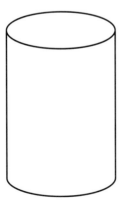 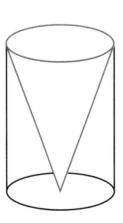 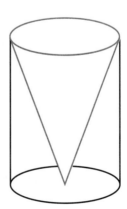 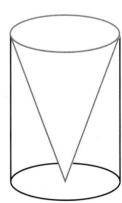

The contents of three conical cups exactly fill a cylinder of the same radius and height.

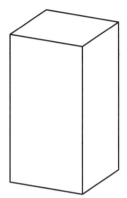 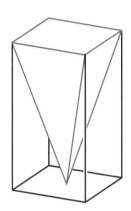 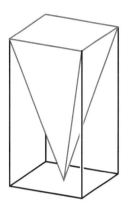 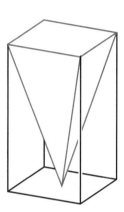

The contents of three pyramids with rectangular bases exactly fill a prism of the same base and height

volume of a cylinder is three times the volume of a cone with like base and height. This relationship was later recognized to hold in a wide range of figures: a prism with rectangular sides and square or triangular base has three times the volume of a pyramid with like base and height. Once we observe this relationship, we can express it in a formula: the volume of a cylinder or prism is the area of the base multiplied by the height, and the volume of a cone or pyramid is one-third the volume of the corresponding cylinder or prism.

The relationship between volumes of prisms and volumes of pyramids is analogous to the relationship between areas of rectangles and areas of triangles in the plane. A diagonal line divides a square into two congruent isosceles right triangles. More generally, the area of a rectangle is the length of the base multiplied by the height, and the area of a triangle is one-half the length of the base multiplied by the height. The pattern lies in the denominator of the fraction. If we consider a four-dimensional analogue of a pyramid, then its four-dimensional volume should be one-fourth the volume of its three-dimensional base multiplied by its height in a fourth direction.

Mathematicians are not content merely to observe a pattern; they want to find an argument that shows why the same pattern will always occur. Experiments with water-filled pyramids do not provide a proof of the volume formula in three dimensions. Fortu-

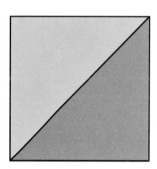

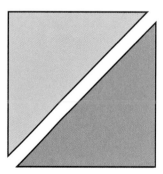

Two congruent right triangles meet along a diagonal of a square.

nately we can establish this relationship by decomposing a cube into three congruent pyramids. Just as a square is decomposed into two congruent triangles which meet along a diagonal line, a cube can be decomposed into three identical pieces which meet along a diagonal of the cube. Each of the pieces is a square-based pyramid, with its top point directly over one corner of the square base. It follows that the volume of each piece is one-third the volume of the cube. From the relationships in the second and third dimensions, the pattern is already clear. We cannot construct a real model to illustrate the analogous relationship in the fourth dimension, but we can still proceed to formulate hypotheses. We expect that a four-dimensional hypercube can be decomposed into four congruent off-center four-dimensional pyramids meeting along the longest diagonal of the hypercube. In general the n-dimensional cube should be decomposable into n congruent off-center n-dimensional pyramids, each with an $(n-1)$-dimensional cube as its base.

The pattern appears both in the exponents and in the denominators of the formulas for area and volume of an off-center pyramid. The area of an isosceles triangle with side length m is $m^2/2$. The volume of an off-center pyramid obtained from a cube of side length m is $m^3/3$. The pattern predicts that the four-dimensional volume of an off-center four-dimensional pyramid obtained from a hypercube of side length m is $m^4/4$. The analogous formula for

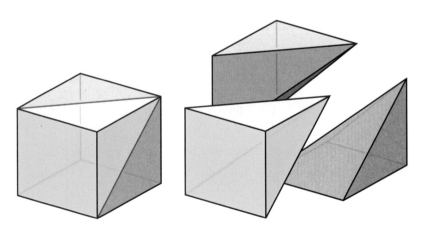

Three congruent pyramids meet along a diagonal of a cube.

the n-dimensional volume of an off-center n-dimensional pyramid is m^n/n.

The analogy between the decomposition of the square in the plane and the cube in space gave us a way to find the volume formula for an off-center pyramid with a square base. We would like to prove a formula for a more general class of pyramids with rectangular bases, but unfortunately the analogy between a rectangular region in the plane and a rectangular prism in space is not a perfect one. A square is divided into two congruent triangles by a diagonal, and so is a rectangular region. But even though a cube can be divided into three congruent pieces meeting along a diagonal, it is usually not possible to decompose a rectangular prism into three congruent pieces meeting along a diagonal. However, it is still possible to divide the rectangular prism into three noncongruent pieces of equal volume, and this result relies on a more sophisticated phenomenon, namely the effect of a scale transformation along a single direction.

We can approximate the volume of a pyramid by stacking thick rectangular cards parallel to the base. If we double the thickness of each card in the stack, then the base stays the same while the height doubles and the weight of the stack (and therefore its volume) also doubles. If we keep the width and the thickness of each card the same and double the length, then the volume also doubles. Doubling any single dimension causes the volume to

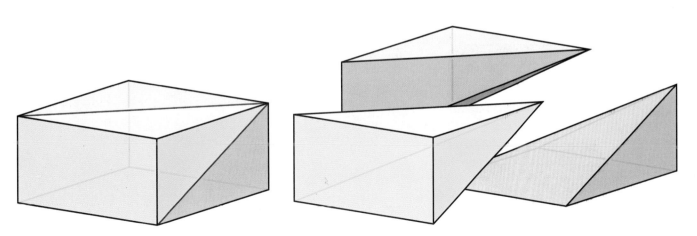

Three noncongruent pyramids with equal volume meet along a diagonal of a rectangular prism.

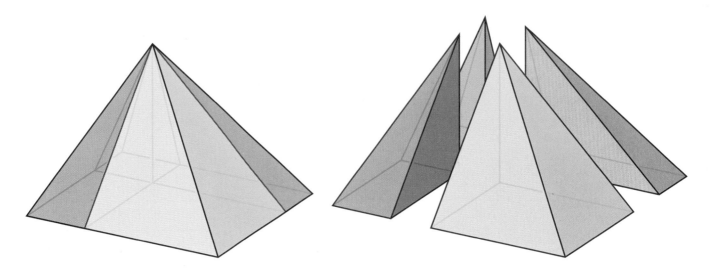

A square-based pyramid is decomposed into four rectangle-based pyramids having their top points over one corner.

double, and in general multiplying a single dimension by any number will multiply the entire volume by that same number.

This procedure enables us to obtain the volume of any pyramid of rectangular base that has its top point directly over a corner of the rectangle. We begin with a cube, having three times the volume of an off-center pyramid contained within it. We double the length of one of the edges of the cube to form a rectangular prism, simultaneously forming an off-center pyramid with the same rectangular base. As we double the length of the edge of the prism, its volume doubles. But all the slabs approximating the volume of the off-center pyramid also double, so the pyramid still has a volume one-third the volume of the prism. The basic volume relationship still holds.

What if the top point of the pyramid is located not over the corner but over some other point of the rectangle? We can then divide the pyramid up into four pyramids with the top point over one corner, and the volume of each of these is one-third the area of the base multiplied by their common height. Adding all of these contributions together shows that in general the volume of a pyramid with a rectangular base is one-third the volume of the rectangular prism with the same base and the same height.

It is also possible to deduce this result by applying Cavalieri's principle for shear transformations, which again uses thick slices to approximate areas and volumes: we can fill a parallelogram with the same set of rods that fills a rectangle of the same base and

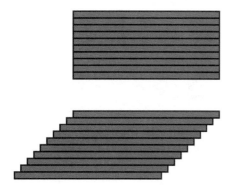

Thin strips that fill a rectangle approximate the area of a parallelogram.

height, and we can fill a slanted box with the same pack of cards that fill a straight box. An off-center pyramid can be approximated with the same collection of square cards that approximate a centered one.

The formulas for volumes of various kinds of square-based pyramids can be modified to give formulas for pyramids formed by cutting off a corner of a cube. The dimension of the space again appears in the denominator of the formula, but in a new way. Cutting off one corner of a square produces a triangle with half the area. Cutting off one corner of a cube produces a pyramid with a triangular base. We can obtain this same triangular pyramid by decomposing a square-based pyramid into four pieces. If we approximate the square-based pyramid by a stack of square slabs, then the triangular pyramid is approximated by a stack of triangular slabs each with half the volume. It follows that the volume of the triangular pyramid is half the volume of the square-based pyramid, or one-third the volume of the triangle-based prism, therefore one-sixth the volume of the cube. The four-dimensional volume of a corner hyperpyramid of a hypercube will be one-fourth the volume of the triangular pyramid which is its three-dimensional base, namely 1/24 the four-dimensional volume of the hypercube. In general the n-dimensional volume of a corner figure of an n-cube will be $1/n!$, where $n!$ stands for "n factorial," obtained by multiplying together the first n numbers.

The Mayan Temple I at Tikal, Guatemala, exhibits the step-pyramid structure used to approximate the volume of an Egyptian pyramid.

The same thin slabs approximate the volume of both a centered square-based pyramid and a pyramid with the same base and its top over one corner.

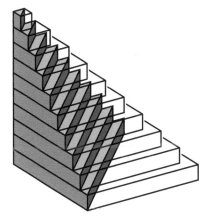

Halving the slabs of a square-based pyramid approximates a triangle-based pyramid.

Fold-out Models of Pyramids

Numerical and algebraic patterns have led us from the familiar geometry of the plane and ordinary space to consider possible objects of higher dimensionality. But how can we begin to understand such objects if we cannot build them in our space? Fortunately the formal patterns suggest analogies in shapes as well as in formulas. By analyzing the way we can construct models of pyramids in three-dimensional space, we can see how to design analogous models for four-dimensional pyramids.

A cardboard model of an Egyptian pyramid can be made by attaching triangles to the sides of a square in the plane and then folding this pattern up into three-space. In the next dimension, we can, in an analogous fashion, erect a square-based pyramid on each of the six square faces of a cube. What we cannot do is fold the pattern together into a fourth dimension to construct the fourth-dimensional analogue of a square pyramid. To find the volume of the three-dimensional Egyptian pyramid that comes from the fold-out model, we can easily calculate the area of the square base, but we still have to figure out how high the top point will be

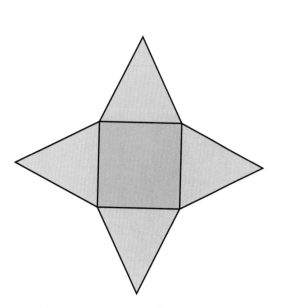

The fold-out pattern for a square-based pyramid.

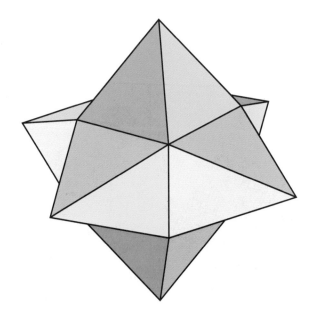

The fold-out polyhedral pattern for a symmetrical cube-based hyperpyramid.

when we fold it up, and this takes some additional calculation. Similarly, we can figure out the volume of the cubical base of the folded-out four-dimensional pyramid, but it is still harder to determine the height in the fourth direction since we cannot actually fold the figure together.

To make a cardboard model of an off-center pyramid, which is one-third of a cube, we start with a square in the plane and at one corner we extend the two sides to obtain two edges of the same length. From these we build isosceles right triangles on two sides of the square. Folding these two triangles together in three-space forms one corner of the off-center pyramid. The other two triangular faces of this pyramid are also right triangles; each has one side along an edge of the square and another side with length equal to the longest side of the isosceles triangle.

An analogue of this construction in the next dimension starts with a cube. At one vertex we extend the three edges to form segments of the same length as the edges of the cube. On these edges we build three off-center pyramids identical to those constructed in the previous paragraph. If we could fold this object up into the next dimension, we would obtain an "off-center hyperpyramid." There are also three more off-center pyramids in the boundary of this object in four-space. Four identical copies of this hyperpyramid would exactly fill a hypercube.

At this point, we are operating only on a formal level, as we do not yet have a means of visualizing what such a decomposition looks like. For that purpose we will develop techniques of projections and fold-outs in later chapters.

The Geometry of the Binomial Theorem

The binomial theorem gives a famous algebraic formula for the sum of two numbers raised to a power. There is a corresponding geometric expression for the volume of an n-dimensional cube with each edge broken into two segments. Earlier in this chapter we considered squares having side length m and area m^2. If we express m as a sum of two numbers p and q, then the algebraic operation of writing m as a sum $p + q$ corresponds to the geometric operation of breaking a segment of length m into two segments of length p and q. If we break a vertical edge of a square having side length m into two segments and draw a horizontal line

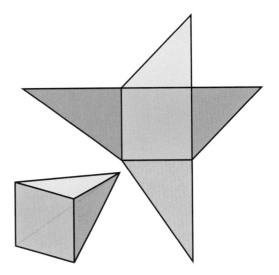

The fold-out pattern for an off-center pyramid.

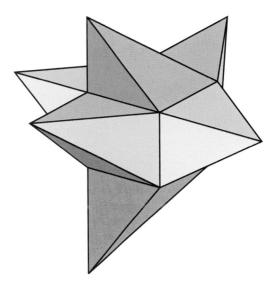

The fold-out polyhedral pattern for an off-center hyperpyramid.

through the break point, we decompose the square into two rectangles. Then doing the same with a horizontal edge produces a
decomposition of the square into four pieces: a square of side
length p, a second square of side length q, and two rectangles
having side lengths p and q. The four parts of this geometric decomposition correspond to the four terms of the expansion of the
square of a binomial. The algebraic problem of finding the square
of expression $p + q$ is equivalent to the geometric problem of finding the area of a square having side length $p + q$.

The rules of algebra enable us to multiply out the square of a
binomial, without having to appeal to a geometric diagram:

$$(p + q)^2 = (p + q)(p + q)$$
$$= p(p + q) + q(p + q)$$
$$= p^2 + pq + qp + q^2$$

We may combine the two terms pq and qp to obtain the familiar
expression for the square of a binomial:

$$(p + q)^2 = p^2 + 2pq + q^2$$

A glance at the diagram below makes the relationship very clear.
Each term of the expression $p^2 + pq + qp + q^2$ gives the area of
one of the four pieces that form the square. The areas of the pieces
add up to the area of the original square, namely $(p + q)^2$.

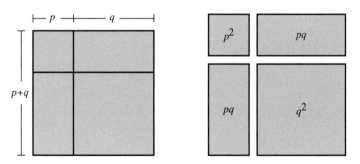

A geometric interpretation of the square of a binomial.

There is a similar pattern relating the volume of a cube having side length $p + q$ in three-space to the third power of the binomial $p + q$. The algebraic procedure is direct enough:

$$(p + q)^3 = (p + q)(p + q)^2 = p(p + q)^2 + q(p + q)^2$$
$$= p(p^2 + 2pq + q^2) + q(p^2 + 2pq + q^2)$$
$$= (p^3 + 2p^2q + pq^2) + (qp^2 + 2pq^2 + q^3)$$
$$= p^3 + 3p^2q + 3pq^2 + q^3$$

Once again, this argument can take place independent of a geometric interpretation, but nonetheless it is useful to give one. A cube having side length $p + q$ can be decomposed into eight pieces: a cube of side length p, another of side length q, and six square prisms, three having height p and square base of side length q, and three having height q and square base of side length p.

What happens to these algebraic and geometric patterns in the fourth dimension? As above, we can obtain the algebraic expression

$$(p + q)^4 = p^4 + 4p^3q + 6p^2q^2 + 4pq^3 + q^4$$

which does not rely on any geometric argument. Nevertheless we can consider a hypercube of side length $p + q$ and describe a geo-

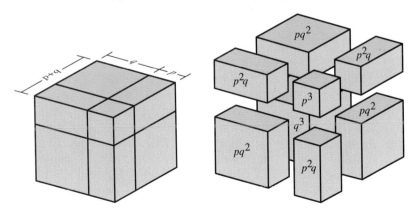

A geometric interpretation of the cube of a binomial.

metric decomposition analogous to the three-dimensional case. This decomposition will have 16 pieces: a hypercube of side length p, another of side length q, and 8 four-dimensional prisms with cubical bases, 4 having height p and cubical base of side length q, and 4 having height q and cubical base of side length p. In addition there are 6 new objects, four-dimensional prisms having at each vertex two edges of length p and two of length q. Thus the geometric pattern generalizes even though we cannot construct in our space a four-dimensional model that will illustrate the decomposition.

The sequences of coefficients that appear in the expansion of binomials are the rows in Pascal's triangle, the famous number pattern that arises in the theory of combinations. Each number in the pattern is the sum of the two numbers above it. We will see these same patterns of numbers occurring in several different guises as we analyze the structures of objects in higher dimensions.

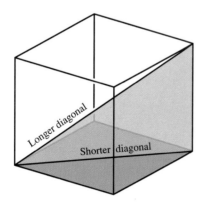

The longer and shorter diagonals of a cube are the hypotenuse and edge of a right triangle.

Diagonals of Cubes in Different Dimensions

We get a different sort of pattern when we analyze the lengths of diagonals of cubes in different dimensions. A diagonal of a cube is a segment joining two points that are not the endpoints of an edge. Whereas both diagonals of a square have the same length, a cube has diagonals of two different lengths, the shorter ones lying on square faces and the longer ones passing through the center. In each dimension, we may compare the length of the side of a cube to the length of its longest diagonal, thus obtaining another significant pattern.

Each time a cube is formed in a new dimension, it means adding edges in a direction perpendicular to all of the directions thus far. The final edge is perpendicular in particular to the longest diagonal of the previous cube, and these two segments form the two sides of a right triangle having the longest diagonal of the new cube as its hypotenuse. This suggests that we can find the length of the longest diagonal by using one of the most famous of all theorems in plane geometry, the Pythagorean theorem.

This theorem states that in a right triangle having sides of length p and q, the length of the hypotenuse is $\sqrt{p^2 + q^2}$. The decomposition we found in interpreting the binomial theorem leads to a proof of this theorem by showing that the area of a square on the hypotenuse of a right triangle is the sum of the areas of the squares on these sides. To see this, we express the area of a square of side length $p + q$ in two ways: once as the sum of the areas of two squares of side lengths p and q and of four right triangles of side length p and q, and again as the sum of the areas of the same four right triangles together with the square bounded by the hypotenuses of those triangles of side length $\sqrt{p^2 + q^2}$.

Applying the Pythagorean theorem to the right triangles formed by cube diagonals sets up an easily discerned pattern. For any dimension, we consider a cube having side length 1. For a square, when $n = 2$, the diagonal has length $\sqrt{2}$. For a cube, when

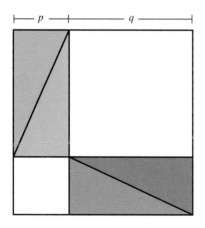
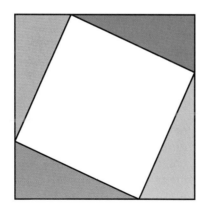

A demonstration of the Pythagorean theorem. Two squares built on the edges of a right triangle and four copies of the triangle exactly fill a large square that is also filled by four copies of the triangle and a square built on the hypotenuse. Take out the triangles, and we have the squares built on two sides of the triangle equal in area to the area of the square built on the hypotenuse.

$n = 3$, the diagonal will be the hypotenuse of a right triangle with base $\sqrt{2}$ and height 1, and by the Pythagorean theorem, the length of the diagonal will be $\sqrt{\sqrt{2}^2 + 1^2} = \sqrt{2 + 1} = \sqrt{3}$. Moving to the four-dimensional hypercube, we obtain a new right triangle with base of length $\sqrt{3}$ and height 1, therefore with hypotenuse $\sqrt{\sqrt{3}^2 + 1^2} = \sqrt{4} = 2$. Thus the hypercube has a diagonal exactly twice the length of a side. It is easy to see that, in general, the length of the longest diagonal of an n-dimensional cube will be $\sqrt{n}$, and this is quickly proved by mathematical induction: if we already know that the length of the diagonal of an $(n - 1)$-cube is $\sqrt{n - 1}$, then the diagonal of the n-cube is the hypotenuse of a right triangle with one side of length $\sqrt{n - 1}$ and the other of length 1. By the Pythagorean theorem, we have $\sqrt{\sqrt{(n - 1)}^2 + 1^2} = \sqrt{n}$ as the length of the diagonal of the n-cube of side length 1.

When we compare a unit cube to a cube of side length m, we observe that the lengths of the diagonals of the new cube are m times as large as the corresponding diagonals of the unit cube. Therefore the length of the longest diagonal of an n-cube of side length m is $m\sqrt{n}$, and the dimension appears under the square root sign. Note that the longest diagonal of a four-cube will be exactly twice as long as one of its edges.

The Egyptian Triumph: The Volume of an Incomplete Pyramid

The rule for finding the volume of a pyramid was of great practical significance to the ancient Egyptians. They could have discovered it easily comparing the amount of sand needed to fill pyramid-shaped and prism-shaped boxes. After filling the pyramid three times with the sand from one prism, they could have realized that the formula for the pyramid is simply one-third times the formula for the prism. Much more difficult to discover by experiment is a formula for finding the volume of an incomplete pyramid, constructed up to a certain level, with the top still to come. How much additional material is necessary to complete the project? Finding such a rule was the most sophisticated discovery of an-

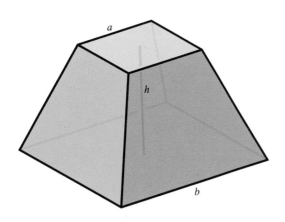

An incomplete pyramid.

cient Egyptian geometry, and it leads to our final example of the ways algebraic and geometric patterns work in higher dimensions.

In order to obtain an intuitive feel and to acquire some important information at the same time, we consider the two-dimensional case of the same problem, the calculation of the area of an incomplete triangle, a trapezoid formed by cutting off the top of a triangle by a line parallel to its base. The key to finding a formula in this case is the principle of similarity. If two triangles are similar, then their bases and their heights are proportional.

We can extend the sides of the trapezoid to complete the triangle. This large triangle is composed of the original trapezoid and a smaller triangle, similar to the larger one. We do not know the height of either triangle, but we know that the ratio of their heights is equal to the ratio of their bases, and this ratio is the same as the ratio of the top and bottom edges of the trapezoid.

The area of a triangle is one-half the base multiplied by the height. If a denotes the base and x the height, then the area is $ax/2$. Analogously, if we have a pyramid of height x and square base of side length a, then the area of the base is a^2 and the volume is $xa^2/3$.

Letting x denote the height of the small triangle, the height of the large triangle is $x + h$. The trapezoid's area is $b(x + h)/2 - ax/2$.

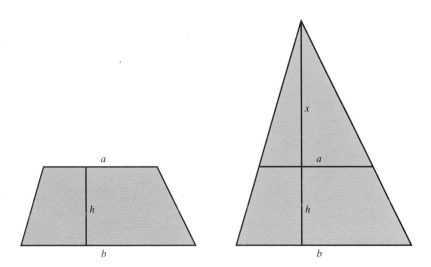

The area of a trapezoid is the difference between the areas of two triangles.

By the similarity principle, $x/a = (x + h)/b = h/(b - a)$. Therefore $x = ah/(b - a)$ and $x + h = bh/(b - a)$, so we obtain

$$\frac{(x + h)b}{2} - \frac{xa}{2} = \frac{hb^2}{2(b - a)} - \frac{ha^2}{2(b - a)}$$

$$= \frac{h(b^2 - a^2)}{2(b - a)}$$

A famous algebraic identity states that $(b^2 - a^2)/(b - a) = b + a$, so the final formula for the area of the trapezoid with height h and parallel sides of length a and b is $h(b + a)/2$.

An analogous method reveals the formula for the volume of the incomplete pyramid. We are given the height h of the incomplete pyramid and the side lengths a and b of the top and the bottom squares. If the height of the large pyramid is $x + h$, then its total volume will be $(x + h)b^2/3$, while the volume of the small pyramid is $xa^2/3$. The argument in the previous paragraph, applied to a vertical slice right through the center of the pyramid, yields $x/a = (x + h)/b = h/(b - a)$, so as before $x = ha/(b - a)$ and $x + h = hb/(b - a)$. Therefore the volume of the incomplete pyramid is

$$\frac{(x + h)b^2}{3} - \frac{xa^2}{3} = \frac{hb^3}{3(b - a)} - \frac{ha^3}{3(b - a)}$$

$$= \frac{h(b^3 - a^3)}{3(b - a)}$$

$$= \frac{h(b^2 + ab + a^2)}{3}$$

The last step is an application of a standard algebraic identity, expressing the difference of two cubes in factored form:

$$b^3 - a^3 = (b^2 + ab + a^2)(b - a)$$

This is the formula known to the ancient Egyptians and described by them in a papyrus dating back to 1800 B.C., using language considerably more complicated than the algebraic formulation given above. The formula represents a high point in the geometry of the ancient world.

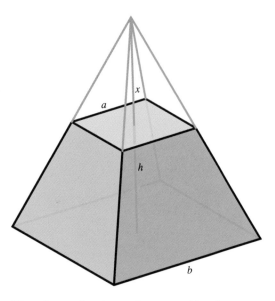

The volume of an incomplete pyramid is the difference between the volumes of two pyramids.

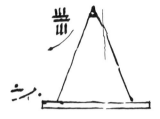
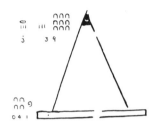

What about the fourth dimension? In a purely formal way, the algebraic pattern suggests that if some four-dimensional artisan constructed a hyperpyramid with height x and a cubical base with edges of length a, then its four-dimensional volume would be $xa^3/4$. The four-dimensional volume of an incomplete hyperpyramid would then be given by

$$\frac{(x+h)b^3}{4} - \frac{xa^3}{4} = \frac{hb^4}{4(b-a)} - \frac{ha^4}{4(b-a)}$$

$$= \frac{h(b^4 - a^4)}{4(b-a)}$$

$$= \frac{h(b^3 + ab^2 + a^2b + a^3)}{4}$$

Problem 57 from the Rhind papyrus deals with areas of triangles and leads to a volume formula for an incomplete pyramid.

We can imagine that a four-dimensional artisan would be as pleased with the discovery of such a formula as the three-dimensional geometer who found the formula for the incomplete pyramid in ancient Egypt.

Scaling and Growth Exponents

When we double the size of an object, the lengths of edges double, the area of square faces goes up by four, and the volume of cubical pieces goes up by a factor of eight. In general we expect that a quantity will increase by a power of two when we double the size, and this growth exponent is the same as the dimension of the quantity. But what of exponents that are not integers? Surprisingly, there are quantities that go up by a power between two and four when their size is doubled. There has been a great interest in recent years in the beautiful images that come from "fractal" objects with noninteger growth exponents.

One of the most popular examples of this phenomenon was invented by the Polish mathematician Wacław Sierpiński. Any triangle can be cut into four congruent triangles, and the first step in creating Sierpiński's figure is to remove the middle triangle. The next step is to remove the middle of each of the remaining triangles. Repeating this over and over again creates what is known as the "Sierpiński gasket."

This object has the remarkable property that doubling its size produces a figure composed of three copies of the original figure. If we double the size of something of dimension one, we get two copies of the original; if we double the size of something of dimension two, we get four copies of the original. The Sierpiński gasket has a dimension such that when we raise two to that exponent, we get three. There is no whole number with this property, and in fact the dimension of the Sierpiński gasket lies somewhere between one and two. Specifically, it is the logarithm of three to the base two.

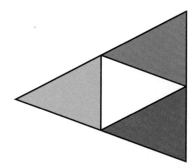

Decomposing a triangle into four congruent triangles.

Six steps in the creation of the Sierpiński gasket.

Another figure that is created by such an infinite process is the Koch snowflake. The snowflake starts with an equilateral triangle. On each of the 3 edges of that triangle, we erect a triangle one-third as large. We then erect a triangle one-ninth as large as the original on each of the 12 edges of the previous figure. Each of these edges is thus replaced by four smaller edges one-third as long as the original. It follows that the total perimeter is multiplied by 4/3 at each stage. Take this process to its limit and we have Koch's snowflake. (In the illustration, we expand the size of the figure at each stage.)

Five steps in the creation of the Koch snowflake.

3 SLICING and CONTOURS

A botanist analyzes a flower bud by embedding it in a plastic cube, then slicing the cube into thin sheets, which she mounts in glass slides. Examining the stack of slides in sequence reveals the interior geometry of the bud. As the botanist goes from the bottom to the top of the stack, she occasionally finds a "critical" slide, where the slice changes form. To get a good idea of the total shape of the bud, she has only to set aside those crucial slides and some representative intermediate slices on either side. Given such a set of slides, it would be possible to re-create a physical model of the bud or to re-create the image in one's mind.

Thanks to recent advances in technology, we no longer have to cut the actual object in order to get a sequence of images of the slices at different heights and in different directions. The first breakthrough was the CAT scan. Computer axial tomography provides X-ray pictures of slices perpendicular to a patient's spine. More recent techniques produce magnetic resonance imaging of a head's interior structure by showing slices not just in the axial direction, perpendicular to the spine, but also from ear to ear, the "sagittal" slices, or from the tip of the nose to the back of the head, the "coronal" slices (see images on the next page). These images show both the bone and the tissue, different shades and textures indicating the densities of various parts. We can mount the images

The contour lines of this terraced hillside in Katmandu can be seen as horizontal slices through the landscape.

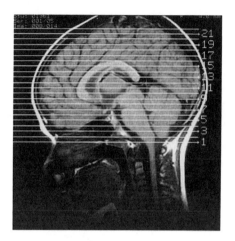 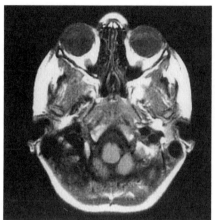 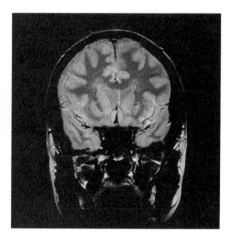

Magnetic resonance imaging produces pictures of the sagittal slice of a head (left), halfway between the ears; the axial slice (center), at the height labeled 1 on the left image; and the coronal slice (right), halfway between the nose and the back of the head.

on plastic sheets, place them in order with the proper separation, and interpolate physically or mentally to obtain a three-dimensional view of the original shape.

In all of these cases, we gain insight into a three-dimensional structure by looking at sequences of two-dimensional slices. If we really understand a solid object, we can predict the slices we will get if we cut it by a series of parallel planes.

A different sort of three-dimensional slide sequence arises when we stack two-dimensional images taken over a period of time. A young amoeba goes about the course of its life unaware that, poised above the petri dish that is its universe, a camera is making a record of its actions, taking a picture every few moments as the amoeba explores its neighborhood and ingests the food it finds. It grows, reaches maturity, and then a day after it first appeared, splits off into two new beings, each event in its life captured by the camera. A technician develops the film onto thin glass squares and stacks them in a tray to form a long prism. Inside that glass prism, we can see a three-dimensional wormlike shape that records the entire life history of the amoeba. To study any event in that history, we have only to select the appropriate slide to get a slice of the amoeba's life.

Edwin Abbott Abbott used slicing to describe communication between different dimensions in *Flatland*. The great climactic event in that story occurs when A Square is visited by a creature from a higher dimension, in this case from our third dimension.

That event totally and irrevocably forces him to revise his understanding of the nature of reality. Imagine A Square as an amoeba-like creature floating on the surface of a still pond, unaware of the air above or the water below, conscious only of the superficial reality, the surface of the water. About to intrude on this two-dimensional universe is a sphere from the third dimension, a beach ball. The water parts as the sphere passes through the surface. To A Square, who can see only the part of the sphere intersecting his plane, the visitation is very mysterious. At each stage he sees only the edge of a plane figure, and he can walk around the figure to observe that it is a perfectly round circle. In Flatland, the circles have all the power, sacred and secular. They are the high priests and philosopher kings. Seeing the passage of the sphere through Flatland, A Square could describe it in only one way. He would say that he had just experienced the accelerated lifetime of a priest! First he would see a priestly zygote, which grows to a circular priestly embryo. A priestly infant is born and grows through minor orders into ordination and full monsignority, only to diminish in size as it reaches old age, until finally it shrinks down to a point and disappears. A Square experiences the successive slices of the sphere as a two-dimensional creature growing and changing in time. In *Flatland*, it takes quite an effort for A Sphere, himself a bit of a mathematician, to convince A Square that A Sphere is not a two-dimensional creature growing and changing in time, but a being that extends spatially beyond the two dimensions of Flatland into a third dimension unknown to the Flatlanders. A Square experiences that third dimension as time, but that does not mean that "the third dimension *is* time."

   And what challenge does that give to us who are privileged to live in Spaceland? Here we are in our three-dimensional "still pond," ready to believe that this is all there is. What would happen if we were visited by a sphere from the fourth dimension? The

Edwin Abbott Abbott in 1884, about the time he wrote *Flatland*.

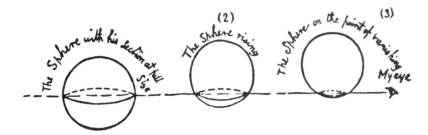

The visitation of A Sphere through Flatland, perceived by A Square as a circle changing in time.

analogy is clear. First we would see a point expanding in every direction we can see to form a small sphere, which grows until it reaches its full distention, then shrinks back down to a point and disappears. That same sequence of growth and decline can be simulated precisely by the gradual inflation and deflation of a balloon. Without more information, it would be impossible for us to tell whether we were seeing an ordinary sphere growing and changing in time or the successive three-dimensional slices of a "hypersphere" from the fourth dimension.

Modern readers are often perplexed by the significance of *time* as a fourth dimension. That is just what it is, *a* fourth dimension, not *the* fourth dimension. Abbott wrote *Flatland* twenty years before anyone had thought of relativity theory, where time is treated as a fourth dimension, so this was not a point of confusion for him, as it is for us who have been brought up in the twentieth century. Certainly people in the nineteenth century were aware that time often appeared in equations and that it could be represented spatially on graphs. People understood and used the idea of time as a fourth coordinate in something as prosaic as making an appointment in a place like New York City. "I'll meet you at the corner of Seventh Avenue and Fourth Street on the fifth floor" might be entered in the calendar as (7, 4, 5), but the appointment would not be complete without that fourth time coordinate, "at ten o'clock," which could be recorded as (7, 4, 5, 10). This four-dimensional system of "three-space, one-time" is immensely useful in modern physics, not just as a bookkeeping device but as a rich mathematical structure. The mathematics is not the same as that of ordinary space though; the time dimension really acts differently. What Abbott is presenting is the challenge of imagining a four-dimensional *space* that is "homogeneous," where every direction is like every other direction, where we can pick up a box and set it down so that no matter which three of the four directions we see, no one is distinguishable from any other.

Curiously, each of the three views used by magnetic resonance imaging has appeared as a device in lower-dimensional science-fiction allegories. *Flatland* encourages us to explore an *axial* view of a two-dimensional universe. In his *An Episode of Flatland*, also written in 1884, C. Howard Hinton, a contemporary and probable inspiration of Abbott, proposed a race of two-dimensional creatures from a *sagittal* viewpoint, right- and left-handed triangular beings living on the outside of a disc. Dionys Burger, in writing his 1964 sequel, *Sphereland*, described symmetrical fig-

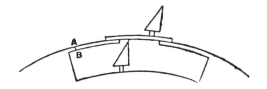

The triangular figures in C. Howard Hinton's *An Episode of Flatland* represent sagittal, or profile, views.

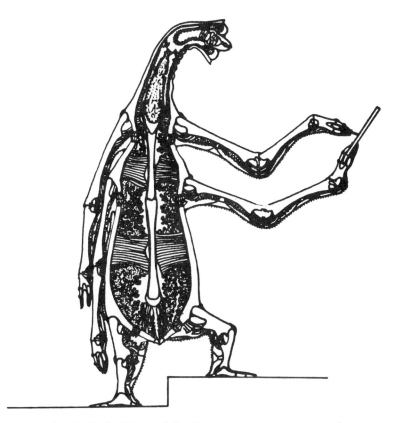

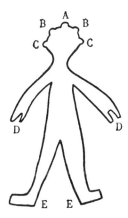

A. K. Dewdney's Yendred, hero of *The Planiverse*, represents a coronal, or straight on, view.

The flat humanoid in Dionys Burger's *Sphereland* is illustrated in a coronal view.

ures in a *coronal* view. In our own day, A. K. Dewdney used a mixture of sagittal and coronal views in his modern allegory, *The Planiverse*. Each approach has its particular geometrical features, and each suggests its own questions.

Slicing Basic Three-Dimensional Shapes

When Abbott wrote *Flatland*, he was certainly aware of the work of Friedrich Froebel, the educational pioneer who invented the term *kindergarten* and who stressed the importance of presenting young children with geometric stimuli. By the 1880s, Froebel's ideas were beginning to influence preschool education in England and

Top: Friedrich Froebel, inventor of kindergarten.
Right: Froebel's geometrical models, from the
Milton Bradley catalogue of 1889.

the United States, as they had in Prussia and other parts of Europe earlier in the nineteenth century.

One of Froebel's first "gifts" to inspire kindergarten children was a display of three basic three-dimensional forms, the sphere, the cylinder, and the cube, suspended by strings. As the objects rotated, children could observe them from different views and learn to appreciate their symmetries and structures.

The objects could be suspended in different ways from eyelets attached to their surfaces. Because of its symmetry, the sphere had only one eyelet, and no matter how it was suspended, it looked the same. The cylinder had three eyelets, one in the center of the top disc, one on the rim, and one halfway down on the side of the cylinder. The cube also had three eyelets, one in the center of a square face, one on the center of an edge, and one at a vertex.

If we suspend the objects from various points, we get different views and, more significantly, different sequences of horizontal slices. We can imagine what will happen if we gradually submerge the set of blocks in a pail of water. How will the shape of the slice formed at the water's surface change? The investigation of different slices of ordinary objects can give us a much better appreciation of their symmetries, as well as the ways their various parts fit together. Such observations will aid us in analyzing more complicated figures in ordinary space, and prepare for the later study of phenomena coming from higher dimensions.

When the sphere goes through the plane, the slicing sequence re-creates the visitation story in *Flatland*: a point, then a small circle, which grows to a large circle, which then shrinks down to a point and disappears. No matter how we suspend the sphere, the story is the same.

Submerging the cube gives more complicated and more interesting results. The three different ways of suspension lead to three very different slicing sequences. We see the least variable series of slices when the cube is suspended from an eyelet in the center of a square face. As the water level rises, all slices are squares, all the same size. If A Square were floating on the surface of the water, he would report that suddenly out of nowhere there had appeared in front of him a square like himself, which remained there for a while, and then abruptly disappeared. He would describe the cube as "a square for a while," mistakenly interpreting one of the dimensions of space as a dimension of time.

How can A Square begin to appreciate what actually happened? In the story, A Square is prepared for understanding such a visitation by a vision of Lineland, a world of only one dimension. The King of Lineland is a long segment, who cannot "see" anyone beyond those two subjects immediately adjacent to him.

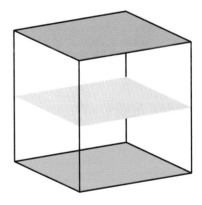

Slicing the cube square first.

A Square appearing to the King of Lineland as "a segment for a while."

When A Square visits this realm, edge first, the King considers him to be "a segment for a while."

What would we see if we were visited by a four-dimensional hypercube? If it came through our space "cube first," we would see "a cube for a while." Our position would be analogous to that of A Square trying to come to terms with a visit of a three-dimensional cube, or the King of Lineland trying to comprehend the passage through his land of A Square.

As it happens, slicing sequences reveal some of an object's important properties. For one thing, we can figure out how many corners or vertices the analogue of a cube has in each dimension. The King knows that as a segment he has 2 vertices, which we might think of as bright points. As he watches the "segment for a while," he sees two bright points at the first instant and two more at the end, so he knows the square has 4 vertices. A Square could appreciate that a cube has 8 corners even though he cannot see a cube all at once in his flat world. A "square for a while" has 4 bright vertices at the first instant and 4 more at the final moment, for a total of 8. Similarly, a hypercube, thought of as "a cube for a while," would have 8 bright vertices at the start and 8 more at the end, for a total of 16 vertices. Even though we can never hope to "see" a hypercube all at once in the same way that we see a cube, we can have confidence that if we ever do see such an object, it will have 16 vertices. If we ever see anything with more vertices or fewer, we will know that it is not a hypercube.

In *Flatland*, we are encouraged not to stop at the second dimension or the third or even the fourth. If we continue the analogy, we can imagine that a five-dimensional cube passing through a four-dimensional "still pond" will appear as "a hypercube for a while," with 32 vertices. Each time we go up a dimension, the number of vertices is multiplied by two. Thus it is that the number of vertices in a square is two times two, or two squared, and the number of vertices of a three-dimensional cube is two cubed. The pattern is clear: in a space of any given number of dimensions, the number of vertices of a cube in that dimension is a power of two, with one factor of two for each dimension. (We may symbolize this by saying that the number of vertices of a cube in n dimensions is 2^n.)

Note that we have not raised the question about whether or not a hypercube actually exists as a physical object. Mathematicians are interested in describing properties of geometric objects whether or not they correspond to any objects that exist in any physical sense. As a mathematical object, a hypercube is an abstraction, but so is a cube or a square. No one has ever seen a perfect square or cube, but nonetheless we can grasp the idea of these shapes. We can also grasp the idea of the hypercube.

Slicing from Other Directions

Coming back to the ordinary cube, we can ask about other ways that it can pass through the water's surface. If it is hung from the center of an edge, then the opposite edge is the first part to get wet. As the cube sinks farther, the slice becomes a rectangle having one pair of sides of length equal to the length of the first edge and another pair of sides of very small length at first. This pair of sides grows until it surpasses the first pair, reaching a maximum length nearly one and a half times that of the original length. Then the length of that pair of sides shrinks back to zero, as the slice becomes a single edge just as the entire cube submerges.

The most difficult slicing sequence to visualize occurs when when we suspend the cube from one of its vertices. The lowest vertex gets wet first and then expands to form a small triangle as the water level meets three of the six square sides. When the cube is almost fully submerged, we see another triangular slice, which shrinks down to a single point, the point of suspension.

You might want to try to imagine what we get in the middle. When the water level has covered exactly one half of the cube, what is the form of the slice? Many people are surprised at the answer. Halfway through the cube, perpendicular to the long diagonal, we get a perfect regular hexagon, with all six sides of equal length and all angles equal. The answer is reasonable. After all, halfway through we expect to hit something, and there isn't any reason to prefer the upper three faces over the lower three, so we should hit all six faces, and in the same way. Therefore just by reasons of symmetry we should expect to see a hexagon. Such an abstract argument is perfectly convincing for many people, but quite a few others prefer to see an illustration. For example, we could fill a transparent cube with colored liquid and hold it up in various positions. Or we could instruct a modern graphics computer to show us the slices in any given direction.

The appearance of the hexagon as the middle slice of a cube tells us something about the symmetry of the cube. The cube has

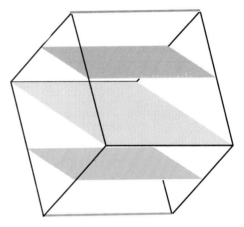

Slicing the cube edge first.

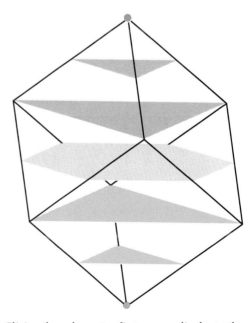

Slicing the cube vertex first, perpendicular to the long diagonal.

A transparent cube half filled with liquid demonstrates the hexagon halfway through the cube.

four long diagonals that pass through its center, and there is a halfway hexagon perpendicular to each of them. The midpoint of each edge will be a corner of two of these hexagons.

Patterns of hexagons and triangles and squares are quite important in the mathematical subject called tiling. Consider what happens if we stack a large number of cubes together to form a large cube. If we slice the large cube by a flat plane, then in each of the cubes hit by the plane, we get a slice in the shape of a polygon. All of these separate pieces will fit together to cover a portion of the slicing plane. If this plane is parallel to one of the faces of the large cube, then all polygonal slices will be squares, and these squares fit together to form a "tiling" pattern in the plane. This is an example of a regular tiling, where all polygons have the same shape. Now consider a slicing plane perpendicular to a long diagonal of the large cube. If such a slice goes through a vertex of a cube, then all of the slices will be equilateral triangles and again we have a regular tiling. If the slicing plane is moved parallel to itself until it goes through the center of a cube, then some of the slices will be regular hexagons, while the adjacent slices will be equilateral triangles. This pattern is a "semi-regular" tiling of the

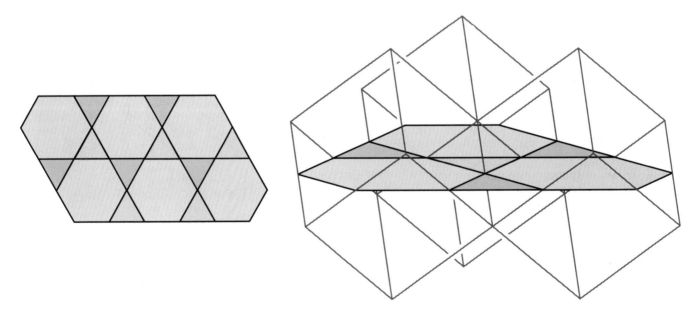

The pattern of triangles and hexagons in a slice through a collection of cubes.

plane, whose polygonal slices are all regular but have varying numbers of sides. Slices in other directions may give tilings of the plane by irregular polygonal figures. The patterns that arise from slicing in different directions are similar to patterns found in crystallography.

Slicing the Hypercube

We have already seen how a hypercube would appear if it penetrated our universe cube first: at the first instant, we would see an ordinary cube with 8 bright vertices, then a dull cube for a while, and finally another cube with 8 bright vertices. What would happen if the hypercube came through our universe from another direction, for example square first or edge first or corner first? For such questions, the modern graphics computer is ideally suited. We can exhibit the slices even though we cannot construct the objects themselves that are being sliced!

We can imagine four-dimensional versions of Froebel's geometrical gifts, a hyperball and a hypercube suspended "above" our three-dimensional world, thought of as the analogue of the surface of the water. How will the appearance of the slices change as the water level rises to cover the objects?

The slicing sequence of the hyperball has already been compared to the inflation and deflation of a spherical balloon. No matter which way the hyperball is rotated, we will get the same sequence. This is analogous to the appearance of the slices of an ordinary ball passing through Flatland. But just as the slices of an ordinary cube depend on the orientation of the cube with respect to the slicing plane, so too do the slices of the hypercube.

If a cube comes through the horizontal plane square first, we see a square for a while, and if the hypercube comes through our space cube first, we see a cube for a while.

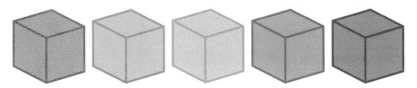

Slices of the hypercube starting with a cube.

A cube coming through edge first was sliced into a series of rectangles. Thinking about the formation of the rectangular slices will guide us to the slices of a hypercube. The squares perpendicular to the leading edge are sliced by a line parallel to one of their diagonals. The slices of each of these squares are line segments: the segments start at a point, grow to a diagonal of the square, and shrink back to a point. Thus the rectangular slices of the cube will have one set of unchanging edges equal in length to an edge of the cube, and another set growing from a point to the length of a diagonal and shrinking back to zero. The 8 vertices of the cube are partitioned into three sets: 2 in the beginning edge, 4 in the largest rectangular slice, and 2 in the final edge.

The comparable slicing sequence for the hypercube starts with a square face, which expands in a series of rectangular prisms all having square bases of the same size. The rectangular sides grow until they reach the length of a diagonal of the original square and then shrink back down to zero. The 16 vertices of the hypercube are thus arranged in three sets: 4 at the beginning, then 8 in the middle, and 4 at the end.

Slicing an ordinary cube corner first yields a point expanding to a triangle, changing to a hexagon, reverting back to a triangle, and shrinking down to a point. The 8 vertices of the cube are arranged in four sets of, in order, 1, 3, 3, and 1 vertices. If we slice a hypercube edge first, we see an edge expanding to a triangular prism with height equal to the length of the original edge. The triangular prism changes to a hexagonal prism, reverts back to a triangular prism, and shrinks down to an edge. The 16 vertices of the hypercube are arranged in four sets of, in order, 2, 6, 6, and 2 vertices.

Finally, if a hypercube comes through vertex first, we start with a single vertex, which expands to form a small triangular pyramid. This tetrahedron expands until it reaches a level con-

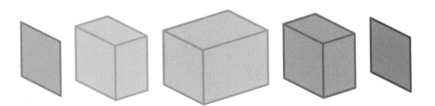

Slices of the hypercube starting with a square.

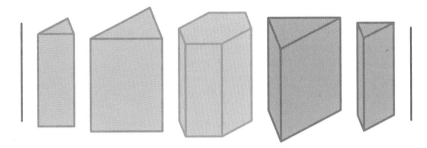

Slices of the hypercube starting with an edge.

taining 4 vertices of the hypercube. Then it experiences truncation as the corners of the tetrahedron are cut off, and three-eighths of the way through, the slice is a solid figure having four equilateral triangles and four regular hexagons as faces. Each hexagon is the slice exactly halfway through one of the cubes in the boundary of the hypercube. This solid is a semi-regular polyhedron known to Archimedes as far back as the third century B.C.

As the slicing plane continues to the position exactly halfway through the hypercube, the four hexagons become triangles, and, together with the previous four triangles, they form a perfect regular octahedron, a Platonic solid that has as its vertices 6 of the vertices of the original hypercube. The second half of the sequence of slices is a reversal of the first half: the original four triangles become hexagons, and the hexagons shrink as the triangles expand until finally there forms a tetrahedron, which shrinks down to the final point of the hypercube. In this final slicing sequence, the 16 vertices of the hypercube are arranged in five dif-

Slices of the hypercube starting with a vertex.

ferent levels: first 1 vertex, then 4 (for a tetrahedron), then 6 (for the octahedron), then 4, and then finally 1.

For a square going through Lineland corner first, the sequence of groups of vertices is 1, 2, 1; for a cube going corner first through Flatland, the vertices appear in successive slices in groups of 1, 3, 3, 1. The corresponding hypercube sequence is 1, 4, 6, 4, 1. This sequence has already appeared as the coefficients in the binomial theorem, and it will arise again when we deal with the coordinates for the hypercube in Chapter 8.

Already in the last century, mathematicians had fashioned models to show the sequence of hypercube slices in various directions. Mrs. Alicia Boole Stott was a genius at predicting the slicing sequences of four-dimensional polyhedra, even though she had no formal training in the techniques of higher-dimensional geometry. In our day we finally have a chance to investigate the objects that our ancestors could only dream about and represent in lifeless models. Interactive computer graphics puts us into direct visual contact with slices of four-dimensional cubes. It is up to us to learn how to interpret these images, and to try to overcome the limitation of our own three-dimensional perspective.

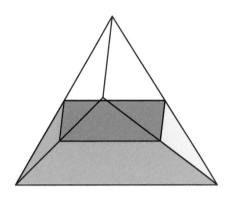

Slicing the tetrahedron edge first produces a square slice in the middle.

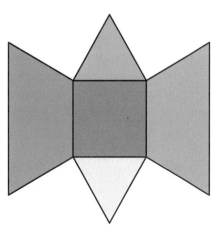

Fold-out pattern for one-half of the subdivided tetrahedron.

Slicing the Triangular Pyramid

We can use these same slicing techniques to investigate figures besides cubes. Consider the slices of a triangular pyramid, or tetrahedron. Slicing parallel to one of the triangular faces, we start with a triangle determined by three of the four points, and these slices become smaller triangles shrinking down to the fourth vertex.

If we slice by planes parallel to one of the edges, we get rectangles, and in the central position, a square. This slicing sequence leads to an interesting two-piece puzzle. The slice containing the square separates the tetrahedron into two parts of exactly the same shape. You can make a paper model of the two pieces of this decomposition by folding up the pattern at the side of this page. Many people find it difficult to put these two identical pieces together to form a triangular pyramid. Even when they place the two square faces together, most often they hold the pieces so that the longest edges are parallel instead of perpendicular as they should be. The difficulty seems to be related to the three-dimensional equivalent of the optical illusion that makes two lines of equal length seem different if we put arrows on the ends. The presence of the longest sides makes the square faces appear to be rectangles with unequal sides.

Slicing Cylinders

Friedrich Froebel's third figure was a circular cylinder. Once we understand the slicing sequences for balls and cubes, it is easy to imagine two of the sequences for the cylinder. If the cylinder is suspended from the center of its circular top, then the slices will all be circular discs. We obtain "a disc for a while," the Flatlanders' perception of a solid cylinder penetrating their world face first.

If we suspend the cylinder from the center of one side so that its circular ends are vertical, then the first slice will be a single segment, which expands to form a rectangle. The rectangle continues to expand until its side length equals the diameter of the circular disc. Then the sequence reverses and the rectangle shrinks back to a single edge.

More complicated are the slices of a cylinder suspended from a point on its rim. Half way through, the slice of the circular cylinder is an ellipse.

Slicing sequences give us a way to think about an analogue of the cylinder in the fourth dimension. We can imagine a figure that in one guise would appear as "a ball for a while," and in another

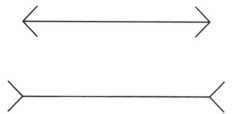

Arrows at the ends of equal-length segments produce the illusion that one is longer.

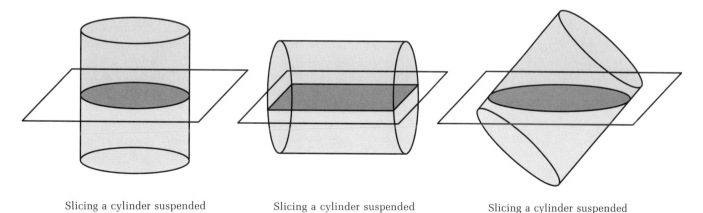

Slicing a cylinder suspended from the top.

Slicing a cylinder suspended from the side.

Slicing a cylinder suspended from the rim.

would appear as a segment, which expands to a thin cylinder, and continues to expand until the cylinder reaches the diameter of the ball, then shrinks back to a single segment. A "diagonal" view of the same object would have in its middle position an ellipsoid, a figure whose plane sections are all ellipses.

Slicing Cones

In the late nineteenth century, a young American draftsman and toy manufacturer, Milton Bradley, took up the task of providing geometric models for kindergartens. To Froebel's trio of sphere, cylinder, and cube, he added a cone, which could also be suspended from different eyelets. Students could imagine the slices, called conic sections, made by various horizontal planes. In presenting this object, Bradley was bringing young students into contact with a distinguished chapter in the history of solid geometry, and he was introducing them to shapes with a great many applications in the physical world.

Apollonius of Perga recognized a fundamental fact that links the second and third dimensions: the ellipse, the parabola, and the hyperbola can all be obtained by slicing a cone in ordinary space. These curves were already important in optics because they gave the shapes of lenses. Geometers knew how to describe these curves as the solutions of locus problems. A *parabola* was the collection of points whose distance to a given *focus* point was equal to the distance to a line lying outside the parabola, called the *directrix*. When the distance from each point to the focus was one-half the distance to the directrix, the collection of points formed an *ellipse*. The same name was given to any curve having the property that the ratio between the distance to the focus and the distance to the directrix was less than one ("ellipse" means "falls short"). A curve having a ratio that is greater than one yields a *hyperbola*.

We see conic sections frequently, for example in the solid cones of light ascending and descending from a lampshade. The boundary of this double solid cone determines the edge of the shadow on any wall in the path of the light. The wall slices the light cone and gives us a conic section.

If we hold a flat plane directly over the shade, we get a circle, growing larger and larger as we move the plane farther and farther

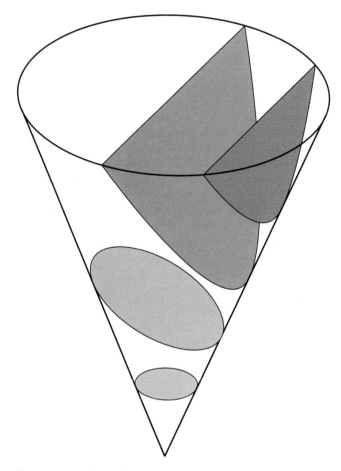

Slicing the cone to produce the conic sections. From top to bottom: hyperbola, parabola, ellipse, circle.

away. If we tilt the plane slowly, the circle of light becomes an ellipse, and moving the plane farther away produces a larger ellipse (having the same ratio between the largest and smallest axis). Tilting farther gives ellipses with greater and greater elongations until finally, as the plane becomes parallel to one of the rays coming past the edge of the shade, the conic section is no longer an ellipse but a parabola, stretching out to infinity.

When we look at a shadow cast by a lampshade, most of the time we observe neither an ellipse nor a parabola. A vertical wall placed next to the lamp cuts out the two branches of a hyperbola, both stretching out to infinity. Frequently the lower and the upper branch are not part of the same hyperbola since the position of the bulb or the slant of the sides of the shade might create an upper cone that does not match the lower cone. If we center a bulb halfway up a cylindrical lampshade, then a vertical wall will slice the light cone to form a complete hyperbola, in theory anyway.

There is a difference between the mathematical discussion and the physical observations that inspired it. In reality a light bulb is not a point source of light, and the edge of a shadow will never be a precise curve. As the slicing plane moves away from the bulb, the image becomes more and more diffuse. To say that an image is a circle or an ellipse is already a mathematical abstraction. To say that an image is a parabola is a much greater abstraction. No matter how intensely the beam were focused, it would take forever to trace out the whole parabola. After all, the beam takes a year to travel just one light-year. Nonetheless, we state without fear of contradiction that in the ideal order the slice of a perfect cone by a perfectly flat plane parallel to one of the lines of the cone will be a perfect parabola. The applications of this to optics, or to planetary motion, are the provinces of physicists and astronomers.

A comet on an elliptical orbit about the sun, such as Halley's comet, will return to our solar system again and again at regular intervals. A comet on a parabolic or hyperbolic orbit will eventually move farther and farther from the sun, finally vanishing from sight never to appear again. After a few observations an astronomer might be able to tell what sort of orbit a given comet is following, although sometimes that is a very delicate proposition. When a comet is on a nearly parabolic path, it is very difficult to tell whether it is on an elliptical path and will return after a great amount of time or on a hyperbolic path with no possible return. Often the shape of the orbit is simply "too close to call."

Contour Lines and Contour Surfaces

Slicing techniques appear in one of the most powerful ways we have for representing three-dimensional information on a two-dimensional page, namely the *contour map*. If we imagine an island completely covered by a flood, then we can generate a contour map by taking an aerial photograph at noon each day, as the water level subsides to reveal the mountaintops and valleys, depressions and mountain passes. For each water level, there is a shoreline, or a collection of different shorelines, all at the same height. We can think of such a shoreline as a horizontal slice of the original island. By keeping track of the way these shorelines change and come together, we arrive at a complete record of the topography of the island. Since each point on the island has exactly one height, no two of these contour lines will intersect. We can then draw all of the lines on the same diagram and number them to indicate the elevation of all of the points on a given contour.

Such a contour map contains a great deal of information about the third dimension, height. An architect could reconstruct a model approximating the surface by cutting out a piece of thick cardboard for the region surrounded by each contour line and stacking the pieces one on top of the other. To get a more accurate model, we could take aerial photographs more frequently to obtain intermediate contour lines, or we could sand off the corners of the model to remove the "terracing." A construction engineer could then weigh the model and compare it to the weight of the cardboard representing a cubic foot of earth in order to estimate the total amount of dirt on the island.

We can gain useful information from the contour map even without constructing a model. To plan her ascent of one of the peaks, a mountain climber can find on the contour map the best way of approaching the base of the mountain across the foothills and can estimate the steepness of various paths to the top.

When we consider the slice history of an island, formed when the flood waters subside, we notice especially the "critical levels" where something interesting happens. A new shoreline suddenly appears as the water level drops below a mountain *peak*, or a lakefront disappears as the water drops below the bottom of a *pit*. (It is best to think of the island as made out of some porous material, so no water gets "trapped" as the level goes down.) Another interesting feature is a *pass*, where two contour lines come together to become one, or when one contour line comes around to meet itself and form two contours. These two cases are exemplified by the slice histories of two islands, which we may call "Twin Peaks" and "Crater Lake."

For Twin Peaks, the first critical level consists of two points, which are the peaks of the two mountains. As the water level subsides, these points grow to small ovals. Then at a third critical

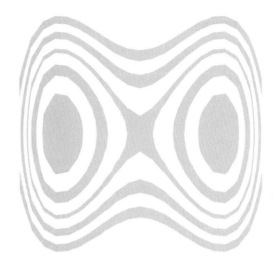

Contour map of Twin Peaks.

Contour map of Tilted Crater Lake.

Slice history of Twin Peaks.

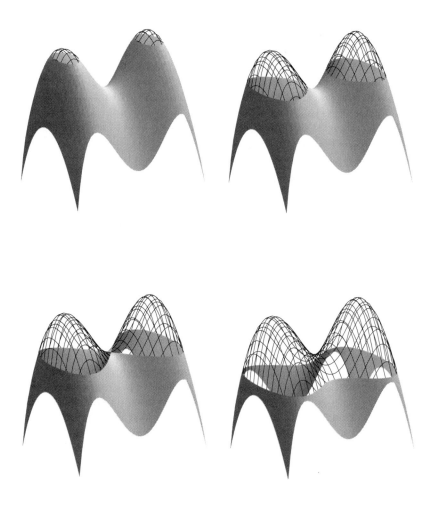

level, the two ovals merge and form a single shoreline. The topology of the island is described by saying that it has two peaks, one pass, and no pits.

Crater Lake has a different sort of slice history. As the water subsides, the first thing we see is a circular rim, which splits into two pieces, a shoreline and a lakefront. The lake then dries up, leaving a single shoreline. If the topography is altered by a small earthquake, which displaces Crater Lake so that one part of its rim is higher than the rest, then we get a slightly different picture. As the water subsides, we see a single point, which grows to an oval. Then two arms reach out from the oval and come together at a

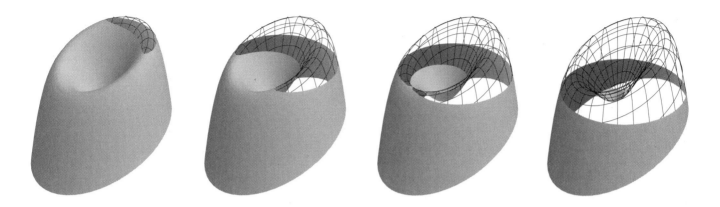

pass, at the low point of the rim. Below this level, we have two contour lines, a shoreline and a lakefront. As before, the water continues to subside, and the lake dries up. The slice history of the tilted Crater Lake shows one peak, one pass, and one pit.

It turns out that, for an island with a single shoreline, the number of peaks plus the number of pits is always one more than the number of passes. This result is called the critical point theorem, and it is the start of one of the most powerful of modern techniques in topology and geometry, with applications to physics and engineering. Generalized critical point theory is called Morse theory, after the American mathematician Marston Morse who extended it to higher dimensions.

We can begin to appreciate the power of critical point theory by placing ourselves once again in the viewing position of A Square, floating on the surface of the water. He himself is unaware that the water level is changing, and he will experience the slice history of Twin Peaks as a sequence of two-dimensional shapes changing in time. As he observes, two ovals appear and then come together to form a single figure. It would be difficult for him to come to a complete appreciation of the actual three-dimensional shape of the island, but at least he could distinguish this island from one having only a single peak. We in three-space do not have to rely only on the sequence of slices since we have the luxury of being able to create a three-dimensional model.

What would be the analogous experience for us, here on the "surface" of our three-dimensional water level? If there were

Slice history of tilted Crater Lake.

Marston Morse and colleague at the dedication of the Institute for Advanced Study at Princeton, 1938.

some hyperisland covered with water, then as the flood waters in the fourth dimension began to subside, two ovaloids might appear and then come together to form a single object. By analogy we realize that we have just observed the slice history of Twin Hyperpeaks, with two high points extending into some fourth spatial direction that we cannot experience directly. This time we do not have the ability to construct the various contours out of cardboard and stack them up to make a model. We have no four-dimensional building materials and no fourth direction in which to stack them up. It is all the more remarkable then that modern computer graphics can exhibit the slice histories of complicated hypersurfaces in four-dimensional space, what the mathematicians call the contours of a three-variable function graph. At the end of Chapter 4, we will use this sort of contour map analysis to help us visualize data sets from geological sciences.

Slicing Doughnuts and Bagels

Slicing techniques are also a tool for investigating smooth surfaces other than spheres and islands. One shape that shows up often enough to merit special consideration is the ring-shaped figure known as a torus. We see this shape in the surface of a doughnut or a bagel, and the surface of a life preserver or a Life Saver candy. By the end of this book, we will encounter the torus in the study of configuration spaces in physics and in the generalizations of perspective to higher dimensions, but for now we wish to consider its slice history as a geometric object in ordinary space

One of the easiest ways to obtain a torus is to generate a surface of revolution. We think of a circle as drawn on a square in a vertical plane, and we attach one vertical edge of the square by hinges to a pole called the axis. As the square rotates around the pole, the moving circle traces out a torus. We can use the same method to generate a sphere, by drawing in the square a semicircle and attaching both endpoints to the axis.

The sphere is called "two-dimensional" because we can identify any point (other than the north and south poles) uniquely by giving two numbers, the latitude showing the position of the point on its semicircle, and the longitude indicating how far the semicircle has been rotated. A torus is a two-dimensional surface in the same sense. We can give latitude and longitude coordinates

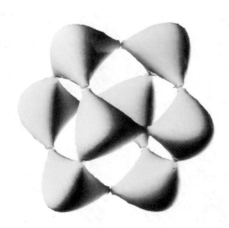

Two hyperplane slices of the graph of a fourth-degree polynomial in four-space. Eight pieces come together at 12 critical points.

for each point on the torus, where now the latitude shows the position anywhere on the vertical circle. Each point on the torus of revolution is specified uniquely by two coordinates. There are no "special points" like the north and south poles on the sphere.

For the slice history of a torus, we think of what happens as we dunk a doughnut in a cup of coffee. The doughnut first meets the surface of the coffee at a single point. If A Square were floating on the surface as the doughnut came through, he would observe the point expand into a small disc, and he might think that he was being visited by a sphere or that he was watching the slice history of an island with a single peak. But something quite different occurs, as two indentations appear on opposite sides of the contour, then come together on the inside as the contour breaks apart into two ovals. Halfway through the surface of the coffee, the doughnut appears as two perfect circles, side by side. The second half of the story is the reverse of the first: two ovals come together and form a single curve, which shrinks to a point as the doughnut disappears below the surface.

Slicing the torus held vertically, as a doughnut.

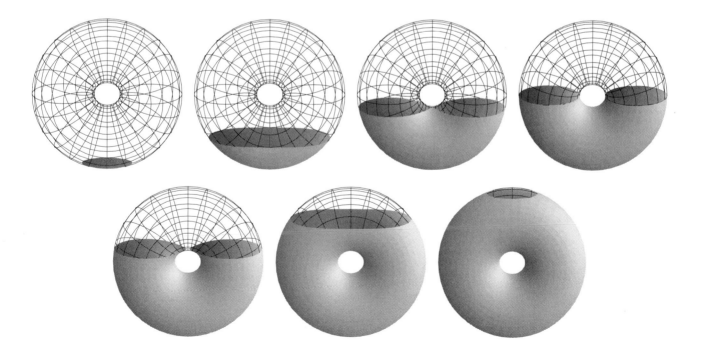

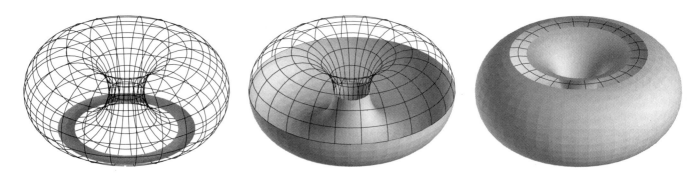

Slicing the torus held horizontally, as a bagel.

There are four critical levels in this slice history, the two points at the top and bottom, and two "figure eights" where pairs of curves break apart or come together. This slice history is quite different from that of a sphere, with its two critical levels each consisting of a single point. Critical point theory gives essential information about the shape of a surface.

Slicing a sphere from different viewpoints gives us no new information since we always get the same sequence of shapes. For a torus, however, different positions tell us quite a bit about the structure of the object. Instead of dunking a torus-doughnut, consider the usual way of slicing a torus-bagel. We set the torus on a plane so that it rests on a circle of latitude. As we slice by horizonal planes, the first slice is a single circle where the bagel rests on the plate. We then get a flat ring-shaped slice with two circular boundaries, each centered at the point where the slicing plane intersects the axis. One circle expands while the other shrinks until we reach the halfway point, after which time the two circles come back together at a top circle. There are only two critical levels, the first and the last circles.

If we tilt the bagel slightly, we get a different phenomenon. Now the slices start with a single point, which grows to be a disc from which two "pseudopods" emerge. These come together at a critical level to form a curve with a loop, similar to the level through the mountain pass in the slice history of the tilted Crater Lake in the previous section. The slice curve then breaks apart to form two closed curves, one inside the other. Halfway through we get a pair of symmetrical ovals, and then the process reverses it-

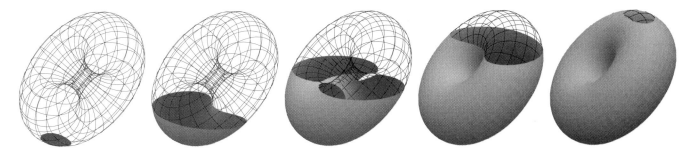

Slicing the tilted torus to reveal two interlocked circles.

self, as the inner oval becomes attached to the outer, then forms a single curve which shrinks down to a point and disappears.

If we continue to tilt the bagel toward a vertical position, its slice history will become the same as that of the doughnut. Somewhere in between there has to be a particularly interesting position where a changeover occurs. At this exceptional position, instead of observing four different critical levels, we see only three. We get a single point at the top and a single point at the bottom as before, but the halfway slice consists of a curve that comes together at two different points, forming two intersecting circles! Each of them goes once around the axis and hits each circle of latitude and each circle of longitude exactly once. The torus is so symmetrical that through every one of its points there pass two of these circles in addition to the circles of latitude and longitude. This remarkable family of circles will appear in an extremely significant way in a later chapter, in the study of orbit spaces of systems of pendulums.

4 SHADOWS and STRUCTURES

Plato was the first person to challenge his readers to contemplate seriously the dimensionality of shadows. He appealed for dimensional empathy to great effect in the seventh book of the *Republic* in his allegory of the cave. In that allegory he proposed an unrealistic scenario: a race of individuals are from birth so constrained in all their movements that they can only see shadows cast on the wall of a cave. They never experience color or shade, and they are never able to touch the objects that cast the shadows. Yet they could obtain quite a bit of information about all sorts of objects from the lower-dimensional representations projected on the wall of their cave. In the allegory, the observers are shown the shadows of different kinds of objects, for example various urns. They could appreciate the shape of an urn, especially if it were turned around so that its symmetry became evident. A tall urn could be distinguished from a short one, and the more discriminating observers of shadows could put together an entire catalog of urns. We who are gifted with sight and freedom of movement can only pity the limitation of these poor creatures. We can imagine the amazement of a prisoner suddenly brought into the open to view the true solid appearance of the objects casting the shadows. The incredible disorientation could well drive the prisoner back to the safe world of the cave. Once the fear was overcome, however, the newly en-

Shadows reflect the forms of realistic and abstract objects in a still life painting by Giorgio Morandi.

lightened individual might have a great desire to share his insights with those still constrained. Plato fully realized the difficulty such a seer would face. The contented cave dwellers would scorn the suggestion that all their hard-earned lore of shadows was inferior to another kind of vision. The prophet could expect only rejection and even persecution, as was the fate of Plato's master, Socrates, described in the *Apology*.

Drawing Shadows

People have long been fascinated by the positions of shadows, as we can easily see by examining the placement of temples in many ancient cultures. Nearly all religious structures were arranged to take account of the position of the sun and of shadows at certain key times of the year. Many of these structures actually functioned as rudimentary observatories, precisely identifying crucial days like the summer or winter solstice by the positions of shadows cast by their monuments. Artisans and astronomers in many cultures devised sundials for precise measurement of time, effectively transforming the passage of time into the movement of a shadow across a plane.

For the most part, shadows are images cast on a plane surface, like a wall or the ground, by some object located between the surface and a light source. In this chapter, we consider shadows cast by the sun, where the light rays are for all practical purposes parallel. (In Chapter 6, we will consider the case where the light rays spread out from a concentrated source, like a candle flame or a laser.) If two parallel lines in space have shadows that are two different lines in a plane, then these shadows are also parallel. This fundamental fact enables us to use shadows to obtain images of complicated structures in three-dimensional space and ultimately in spaces of higher dimension.

Drawing Cubes and Hypercubes

One way to draw a cube is to build one out of sticks in three-dimensional space and hold it up in the sunlight. The edges of the cube cast shadow lines, which we can trace on a piece of paper. Such a procedure is awkward, but fortunately we do not have to trace the entire object, just enough to enable us to complete the drawing at our leisure. An easy way to do this is to sketch the images of three lines emanating from one corner of the cube and then complete the image by drawing three sets of four parallel lines.

Since parallel lines in space have images that are parallel lines in the plane, the image of a parallelogram is always a parallelogram (possibly degenerating into a line segment if the plane of the parallelogram is parallel to the rays of the sun). Thus if we know the images of two edges of a square, we can easily complete the image of the other two edges.

Once we are given just the images of the edges from one corner, the process of completing a picture of a cube or a square is so straightforward that we can tell a computer how to do it. Even a small computer can draw several such pictures in a second, and it is now a standard demonstration on many machines to create a sequence of images of a cube, one after another, each differing only slightly from the previous one, to form a real-time animated cartoon. The information given to the computer is exactly equivalent to the position of three edges coming from a vertex of a cube. For the sake of the computer, this information is transmitted in numerical form rather than in pictures. The position of the vertex on the computer screen is given by a pair of numbers, and each of the endpoints of the three segments coming from the vertex is given by another number pair. Thus it is that eight numbers in a particular order give the precise information that allows the computer to display the three edges and then to figure out the location of all other edges of the cube and display them as well. We will treat these numerical descriptions further in Chapter 8.

The beautiful thing about this method of describing the views of a cube is that we never have to build the cube in the first place. We could use the same method to design a thousand-story building, and with some additional techniques, we could show the shape of the shadow it would cast at noon on a certain day at a certain latitude. How much more practical it is to use this abstract method than to build the object and trace the shadow!

A dramatic example of this method is the story of the Johnson Art Museum at Cornell University. Cornell University has been a pioneer in the application of computer graphics to architectural design, so it was natural for the campus planners to seek advice from people in the Program of Computer Graphics in the College of Architecture, Art, and Planning. The planners agreed that the

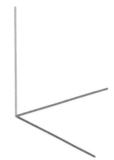

The three edges at the vertex of the cube.

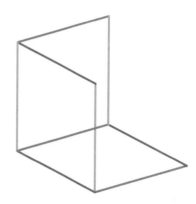

Completing three squares at the vertex of the cube.

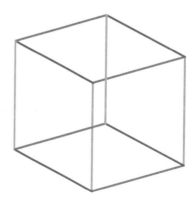

Adding three edges to complete the cube.

This simulated image of a contemporary house
was rendered using a program that indicates the
position of shadows. The synthetic environment
was superimposed on a scanned photograph.

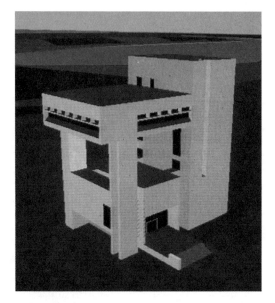

This image of the Johnson Art Museum at Cornell
University was created in 1970 at the Visual
Simulation Laboratory of the General Electric
Corporation in Syracuse, New York.

multistory building should be erected close to the centers of activity. The precise location, however, depended on several factors,
primarily the desire that the new structure literally not over-
shadow some of the traditional buildings or disrupt the balance of
the Arts Quadrangle. The computer graphics experts under the
direction of Donald Greenberg were able to answer these questions definitively. An interactive program gave the designers the
option of selecting a site for the building, then seeing precisely
how its presence affected the surrounding buildings. Armed with
this information, the building committee made their selection
confidently. An image from this project became the May, 1974
cover for *Scientific American*.

Shadows of Hypercubes

Using computer graphics, we can analyze the shadows of a three-
dimensional building even before it is constructed. And we can
do more. We can investigate the shadows of objects that never
could be built with our three-dimensional materials, for example

the four-dimensional hypercube. We can draw a hypercube on the page of a book the same way we drew the two- and three-dimensional cubes, by drawing the images of the edges coming out of one corner and then drawing a collection of parallel edges to complete the figure. Two edges emanating from a point determine the image of a square, and three edges determine the image of a cube. Four edges emanating from a point will determine the image of a four-dimensional cube. First taking two edges at a time, we complete six parallelograms; then taking three edges at a time, we complete the images of four different cubes. Drawing the remaining four edges will complete the figure.

We can generalize this process to even higher dimensions since each dimension has its analogue of the cube. If we draw five edges from a corner, we can complete the image of a five-dimensional cube, or five-cube. There is nothing to stop us from drawing shadows of hypercubes of any number of dimensions on a two-dimensional paper or a two-dimensional computer screen.

We can't construct a hypercube in four-dimensional space and draw its actual shadow, but we don't have to. We know that if such a cube existed and if it cast a shadow on a wall, then we would have four groups of eight parallel edges. If the initial set of four edges changed slightly, then the whole figure would follow along, and by recording a succession of slightly changed images, we would obtain an animated cartoon of the transforming object. Naturally this process would be tedious to do by hand. Already in the last century mathematicians and draftspersons sketched individual shadows of this type, but it took the development of modern graphics computers to produce animated films of rotating hypercubes. The first such films were made by A. Michael Noll and his associates at Bell Laboratories in the 1960s. The most complete version, *The Hypercube: Projections and Slicing*, was produced by Charles Strauss and the author at Brown University in 1978 as part of an invited presentation at the International Congress of Mathematicians in Helsinki.

Three-Dimensional Shadows of the Hypercube

Reasoning by analogy, we can imagine a sun in four-dimensional space casting a three-dimensional shadow of a hypercube. To make such a shadow in three dimensions, we can proceed as we

Starting from four edges, six squares are drawn at the vertex of the hypercube.

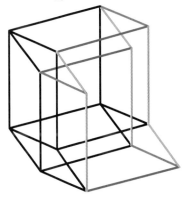

Adding the rest of the four cubes at the vertex of the hypercube.

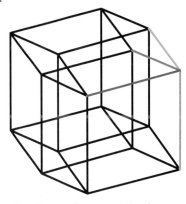

The complete hypercube, containing four sets of parallel edges.

did when drawing shadows in the plane. We start with four edges coming out of a vertex, no three of which lie in the same plane. We complete the pairs of edges to form six parallelograms, then complete triples of edges to form four *parallelepipeds*, distorted cubes with all six sides parallelograms. Finally, we put in the last four edges to obtain the four groups of eight parallel edges each. We can make such a model from sticks or wire, or we can instruct a computer to show us what such a model would look like if we filmed it rotating around in three-space. Even though the images on the computer screen are two-dimensional, the computer can produce animated sequences simulating the form of three-dimensional shadows of higher-dimensional cubes.

Even before the emergence of graphics computers, artists and designers were constructing three-dimensional images of objects from the fourth and higher dimensions. Two remarkable examples come from nonmathematicians who became fascinated by the challenge of visualizing these strange objects. The wire models of projections of higher-dimensional cubes and other objects built by Paul R. Donchian are part of a permanent exhibit at the Franklin Institute in Philadelphia. (One model is shown in Chapter 5.) David Brisson, professor at the Rhode Island School of Design and founder of the Hypergraphics group of artists, produced sculptures of cubes in four, five, and six dimensions. (A watercolor of two views of a hypercube is shown in Chapter 6.)

Counting the Edges of Higher-Dimensional Cubes

On first view, a hypercube in the plane can be a confusing pattern of lines. Images of cubes from still higher dimensions become almost kaleidoscopic. One way to appreciate the structure of such objects is to analyze lower-dimensional building blocks.

We know that a square has 4 vertices, 4 edges, and 1 square face. We can build a model of a cube and count its 8 vertices, 12 edges, and 6 squares. We know that a four-dimensional hypercube has 16 vertices, but how many edges and squares and cubes does it contain? Shadow projections will help answer these questions, by showing patterns that lead us to formulas for the number of edges and squares in a cube of any dimension whatsoever.

It is helpful to think of cubes as generated by lower-dimensional cubes in motion. A point in motion generates a segment; a

The 1983 work *Simplex* of Tony Robbin combines painting and wire sculpture to present geometric forms and their shadows in different dimensions.

segment in motion generates a square; a square in motion generates a cube; and so on. From this progression, a pattern develops, which we can exploit to predict the numbers of vertices and edges.

Each time we move a cube to generate a cube in the next higher dimension, the number of vertices doubles. That is easy to see since we have an initial position and a final position, each with the same number of vertices. Using this information we can infer an explicit formula for the number of vertices of a cube in any dimension, namely 2 raised to that power.

What about the number of edges? A square has 4 edges, and as it moves from one position to the other, each of its 4 vertices traces out an edge. Thus we have 4 edges on the initial square, 4 on the final square, and 4 traced out by the moving vertices for a total of 12. That basic pattern repeats itself. If we move a figure in a straight line, then the number of edges in the new figure is twice the original number of edges plus the number of moving vertices. Thus the number of edges in a four-cube is 2 times 12 plus 8 for a total of 32. Similarly we find $32 + 32 + 16 = 80$ edges on a five-cube and $80 + 80 + 32 = 192$ edges on a six-cube.

By working our way up the ladder, we find the number of edges for a cube of any dimension. If we very much wanted to know the number of edges of an 11-dimensional cube, we could carry out the procedure for 10 steps, but it would be rather tedious, and even more tedious if we wanted the number of edges of a cube of dimension 101. Fortunately we do not have to trudge through all of these steps because we can find an explicit formula for the number of edges of a cube of any given dimension.

One way to arrive at the formula is to look at the sequence of numbers we have generated arranged in a table.

	Dimension of cube					
	1	2	3	4	5	6
Number of vertices	2	4	8	16	32	64
Number of edges	1	4	12	32	80	192

If we factor the numbers in the last row, we notice that the fifth number, 80, is divisible by 5, and the third number, 12, is divisible by 3. In fact, we find that the number of edges in a given dimension is divisible by that dimension.

	Dimension of cube					
	1	2	3	4	5	6
Number of edges	1	2×2	3×4	4×8	5×16	6×32

This presentation definitely suggests a pattern, namely that the number of edges of a hypercube of a given dimension is the dimension multiplied by half the number of vertices in that dimension. Once we notice a pattern like this, it can be proved to hold in all dimensions by mathematical induction.

There is another way to determine the number of edges of a cube in any dimension. By means of a general counting argument, we can find the number of edges without having to recognize a pattern. Consider first a three-dimensional cube. At each vertex there are 3 edges, and since the cube has 8 vertices, we can multiply these numbers to give 24 edges in all. But this procedure counts each edge twice, once for each of its vertices. Therefore the correct number of edges is 12, or three times half the number of vertices. The same procedure works for the four-dimensional cube. Four edges emanate from each of the 16 vertices, for a total of 64, which is twice the number of edges in the four-cube.

In general, if we want to count the total number of edges of a cube of a certain dimension, we observe that the number of edges from each vertex is equal to the dimension of the cube n, and the total number of vertices is 2 raised to that dimension, or 2^n. Multiplying these numbers together gives $n \times 2^n$, but this counts every edge twice, once for each of its endpoints. It follows that the correct number of edges of a cube of dimension n is half of this number, or $n \times 2^{n-1}$. Thus the number of vertices of a seven-cube is $2^7 = 128$, while the number of edges in a seven-cube is $7 \times 2^6 = 7 \times 64 = 448$.

Higher-Dimensional Simplexes

The shadow image of an Egyptian pyramid is completely determined by the shadow of its apex. At any instant, the shadow of an edge along the base remains fixed, and the image of an edge leading from a base point to the apex will be a line joining the base point to the image of the apex. This fact is true for a pyramid with any shape base. For a triangular pyramid, as soon as we know the location of the shadows of the four vertices, we can draw the complete shadow of the transparent object—just connect all possible pairs of vertices. Thus we can complete the shadow without ever having to construct the object.

The triangle and the triangular pyramid have higher-dimensional analogues, known as *simplexes*. Three points in a plane, not lying in a line, determine a triangle, also called a two-simplex. Four points in space, not lying in a plane, determine a triangular pyramid, also called a three-simplex or tetrahedron. The n-simplex is the smallest figure that contains $n + 1$ given points in n-dimensional space and that does not lie in a space of lower dimension. If we analyze the numbers of edges, triangles, and other simplexes in an n-simplex, we find a pattern that occurs over and over again in algebra and in the study of probability. The reappearance of such patterns is one of the most beautiful aspects of mathematics.

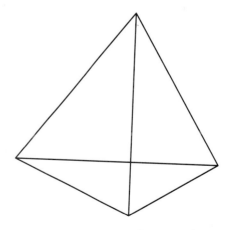

A shadow of a triangular pyramid.

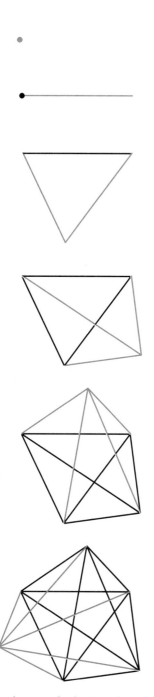

The simplexes in the first six dimensions.

As we follow the procedure for drawing simplexes, we can count the number of edges. We start with a point. We then choose a distinct point and draw the 1 edge connecting it to the point we already had. We choose a new point, not on the line of the first 2 points, and connect it to the previous 2 points to get 2 more edges for a total of 3. The next step is to choose a new point not lying on any of the 3 lines determined by the edges already constructed, and we then connect this new point to the previous 3 points to get 3 new edges for a total of 6.

We repeat this process to draw the image of the simplest figure determined by 5 points, the four-simplex. We first choose a point not on any of the 6 lines containing previously constructed edges and then connect this point to the previous 4 points to obtain 4 new edges for a total of 10. We can imagine that the fifth point actually extends into a fourth dimension, and that we are observing the shadows of the segments connecting it to the 4 points of the three-simplex.

If we arrange the results in a table, a pattern emerges.

	Dimension of simplex					
	0	1	2	3	4	5
Number of vertices	1	2	3	4	5	6
Number of edges	0	1	3	6	10	15

The number of edges at any stage is the sum of all the numbers less than the number of that stage, so the number of edges formed by 6 points will be $1 + 2 + 3 + 4 + 5 = 15$. This can easily be seen from our procedure since each new point is connected to all the previous points.

The study of combinations provides a different way of finding the number of edges of a simplex. Since each vertex is to be connected to every other vertex, the number of edges will be equal to the number of pairs of vertices, that is, the number of combinations of a certain number of vertices taken two at a time. If we have $n + 1$ vertices, then there are $n + 1$ choices for the first element of the pair and n remaining vertices for the second element. Multiplying these numbers gives $(n + 1)$ times n, and since this counts each edge twice, the total number of edges is $n(n + 1)/2$.

Spatial perception tests are fond of asking students to extract a simple figure from a complicated one, and counting the edges is one of the most basic tasks. Next in difficulty would be counting the number of distinct triangles at each stage. The three-dimensional simplex has 4 triangles. The four-dimensional simplex has these 4 plus 6 new triangles formed by attaching the edges of the three-simplex to the new vertex, yielding a total of 10 triangles on the four-simplex. We can then extend the previous table.

	Dimension of simplex					
	0	1	2	3	4	5
Number of triangles	0	0	1	4	10	?

At each stage, the number of triangles formed by a given number of points is the sum of the previous number of triangles and the previous number of edges. Thus the number of triangles in a five-dimensional simplex determined by 6 points is $10 + 10 = 20$. In general, the number of triangles formed by n points will be $n(n - 1)(n - 2)/6$. We have as many triangles as there are distinct

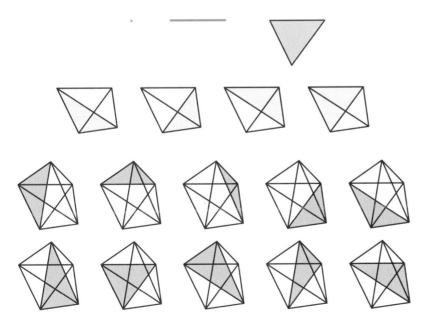

The triangular faces of the simplexes in the first five dimensions.

triples of vertices, so we have the combinations of a certain number of objects taken three at a time. In general the number of k-dimensional simplexes in an n-dimensional simplex is the number of combinations of $n + 1$ elements taken $k + 1$ at a time. The formula for this number is

$$C(n + 1, k + 1) = \frac{(n + 1)!}{(k + 1)!(n - k)!}$$

where $n!$ stands for the product of the integers from 1 to n. The number of k-dimensional simplexes that arise in this analysis appear as the coefficients in the powers of binomials already treated in Chapter 2. They arise in a different way when we attempt to count the numbers of faces of cubes in different dimensions.

Counting the Faces of Higher-Dimensional Cubes

Analogous to the sequence of simplexes in each dimension, we have a sequence of cubes. We begin a table.

	Dimension of n-cube				
	0	1	2	3	4
Number of 0-cubes	1	2	4	8	16
Number of 1-cubes	0	1	4	12	?
Number of 2-cubes	0	0	1	6	?
Number of 3-cubes	0	0	0	1	?
Number of 4-cubes	0	0	0	0	1

When we try to fill in the missing numbers for a hypercube, the process becomes a bit more difficult. We know that we can generate a hypercube by taking an ordinary cube and moving it in a direction perpendicular to itself. We can show what is happening schematically by drawing two cubes, one obtained by displacement from the other. We draw the first cube in red and the second in blue. As the red cube moves toward the blue cube, the

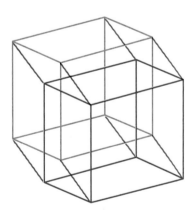

Moving a cube perpendicular to itself creates a hypercube.

8 vertices trace out 8 parallel edges. We have 12 edges on the red cube, 12 on the blue, and now 8 new edges for a total of 32 edges on the hypercube.

Finding the number of square faces on the hypercube presents more of a problem, but a version of the same method can solve it. There are 6 squares on the red cube and 6 on the blue one, and we also find 12 squares traced out by the edges of the moving cube for a total of 24.

The edges in the hypercube come in four groups of 8 parallel edges. Similarly the squares can be considered as six groups of 4 parallel squares, one such square through each vertex. The illustration on the bottom, left, shows two groups of 4 parallel squares. Another group is shown on the right. We can then go on to identify the remaining three groups of 4 squares to obtain the entire set of 24 squares in the hypercube. Note that it is easier to identify the 4 squares when they do not overlap and relatively more difficult when the overlap is large.

This manner of grouping the faces of an object is particularly effective when the object possesses a great deal of symmetry, as does the hypercube. A segment possesses one symmetry, obtained by interchanging its endpoints. A square has a much larger number of symmetries: we can rotate the square into itself by one, two, or three quarter-turns about its center, and we can reflect the square across either of its diagonals, or across the horizontal or vertical lines through its center. The even larger group of symmetries of the cube enables us to move any vertex to any other vertex

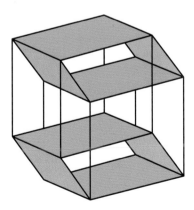

Two groups of four parallel square faces in a hypercube.

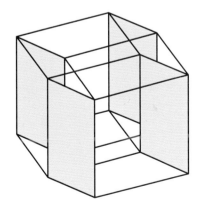

One more group of four parallel faces in a hypercube.

and any edge and square at that vertex to a chosen edge and square at the new vertex. The collection of symmetries is one of the most important examples of an algebraic structure known as a group. The analysis of symmetry groups has provided extremely significant tools in modern geometry and in the applications of geometry to molecular chemistry and quantum physics.

The hypercube is so highly symmetric that every vertex looks like every other vertex. If we know what happens at one vertex, we can figure out what is going to happen at all vertices. At each vertex there are as many square faces as there are ways to choose 2 edges from among the 4 edges at the point, namely 6. Since there are 16 vertices, we can multiply 6 by 16 to get 96, but this counts each square four times, once for each of its vertices. The correct number of squares in a hypercube is then 96/4, or 24.

It is possible to express these results in a general formula. Let $Q(k, n)$ denote the number of k-cubes in an n-cube. To calculate $Q(k, n)$ we may first find out how many k-cubes there are at each vertex. There are n edges emanating from each vertex, and we get a k-cube for any subset of k distinct edges from among these n edges. Therefore the number of k-cubes at each vertex of an n-cube is $C(k, n) = n!/(k!(n - k)!)$, the combinations of n things taken k at a time. Since we have $C(k, n)$ k-cubes at each of the 2^n vertices, we obtain a total number $2^n C(k, n)$. But in this count, each k-cube is counted 2^k times, so we divide by that number to get the final formula: $Q(k, n) = 2^{n-k} C(k, n)$.

An inspection of the table reveals that the entries in each column add up to a power of 3.

	Dimension of n-cube				
	0	1	2	3	4
Number of 0-cubes	1	2	4	8	16
Number of 1-cubes	0	1	4	12	32
Number of 2-cubes	0	0	1	6	24
Number of 3-cubes	0	0	0	1	8
Number of 4-cubes	0	0	0	0	1
Sum of k-cubes	1	3	9	27	81

There are several ways to verify this observation. We can observe that each entry is the sum of twice the entry directly to the left of it plus the entry above that one, so the sum of entries in one column is three times the sum of the entries in the previous column. This argument can easily be translated into a formal proof by mathematical induction. We may also use the explicit formula for the number of k-cubes in an n-cube and observe the sum

$$Q(0, n) + Q(1, n) + \cdots + Q(n - 1, n) + Q(n, n)$$

$$= 2^n + C(1, n)2^{n-1} + C(2, n)2^{n-2} + \cdots + C(n - 1, n)2 + C(n, n)$$

$$= (2 + 1)^n = 3^n$$

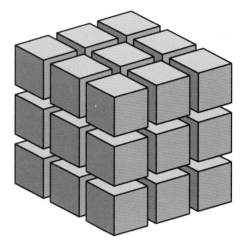

A cube subdivided into 27 smaller cubes.

The most satisfying demonstration of this algebraic fact comes from the observation that we may divide the sides of an n-cube into three parts and extend this to divide the entire cube into 3^n small cubes. We then have a small cube coming from each of the vertices of the original, and one from each edge and one from each two-dimensional face, and so on. The final small cube is in the center. Thus the total number of small cubes is the sum of the numbers of k-cubes in the n-cube, and this value is 3^n.

One of Friedrich Froebel's kindergarten gifts was a cube subdivided into 27 small cubes. He would have liked this final demonstration.

Paleoecology and Data Visualization

The techniques of slicing and projection are powerful tools in analyzing geometric figures and also in visualizing complicated data sets. We illustrate the application of these techniques to data visualization by presenting an extended example that uses many of the topics of this chapter and the previous one.

For a major research project in paleoecology, Professor Tom Webb and his co-workers in the Geological Sciences Department at Brown University are studying climate changes over thousands of years by tracking the changing vegetation. Their typical data come from counting pollen grains in cores of lake sediments. Within these long vertical cylinders of packed soil, the pollen grains are preserved in temporal order, the oldest material lying in the lower part of the core and the more recent material lying near

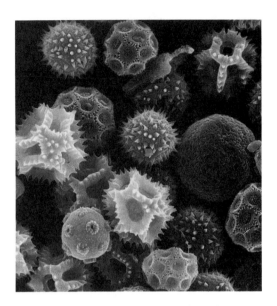

Microscopic grains of fossilized pollen. The different varieties have distinct appearances.

the top. To verify the chronology, the sediments are radiocarbon dated. By examining the abundance of different types of pollen, the researchers can tell the distribution of different varieties of trees, herbs, and grasses. In particular, they can tell whether the land was forest or prairie by comparing the amount of oak or spruce pollen to the amount of forb pollen from prairie grass.

Traditional studies in this field have usually analyzed readings at a single site. Researchers count the pollen grains in different layers of a sample core, knowing that the deeper down in the core, the further back the layers go into the history of the site. The abundance of forb pollen at the site is plotted in a time series, a graph where the horizontal axis indicates the time when the pollen was deposited and the vertical axis gives the percentage of pollen. This visual presentation of the data can show how the abundance rises and falls over periods of thousands of years. But this single time series cannot show how changes at this particular site are related to the changes at nearby sites. Perhaps we happened to choose an isolated lake where a bog was forming, which induced vegetational changes not at all representative of the broad area surrounding it.

To eliminate this possibility, we could compare the time-series information from one site with that from another nearby site. If we drew the graphs on transparent sheets, we could overlay the two graphs to see where changes in one site are located to the right of those in the other, indicating that the changes occurred later in time, and in general we could compare the relative shapes of the graphs to see if they exhibit the same overall behavior. We could even take the difference of the values in the two graphs to make even clearer the places where the two readings differ significantly, a process that goes under the name high-frequency filtering.

If we have a sequence of sites along a transect, for example a path up a mountain slope or along some arbitrary boundary line like a circle of latitude, then we can stack up the graphs corresponding to all of these sites and get an idea of the changes in vegetation along that entire transect. The collection of transparent sheets displaying the graphs will define a three-dimensional "viewing box" with one axis representing time (or depth of the core), another representing space (distance along the particular path), and the third giving the abundance of one or more types of pollen. If we color code the different varieties of pollen, then we can display several of them on the same three-dimensional dia-

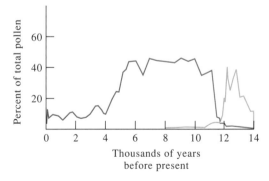

A time series shows the changing percentages of pollen in forb and pine at one site in Michigan over 14,000 years.

gram. As before, we could take the difference of two of these graphs to exhibit more clearly where the abundance of forb pollen exceeds, for example, that of spruce pollen. The higher dimensionality of the display enables us to deal with more data simultaneously and to see relationships that might not be apparent in tabular displays of the data or in isolated time series.

But the data set in reality is of even higher dimensionality. Sites are spread over an entire region, not just up a mountainside or along a transect. We have a two-dimensional region with a two-dimensional time series of forb abundance versus time at each site within the region. The data set is four-dimensional. How do we deal with such a collection?

Our earlier experience with mathematical constructs of higher dimensionality gives us some clues for analyzing such aggregates: we find ways of reducing the dimensionality by slicing or projecting. Slicing is the most immediately useful tool in this sort of investigation. By stacking up the graphs corresponding to the points along one parallel of latitude, we obtain a three-dimensional slice of the four-dimensional configuration. A sequence of

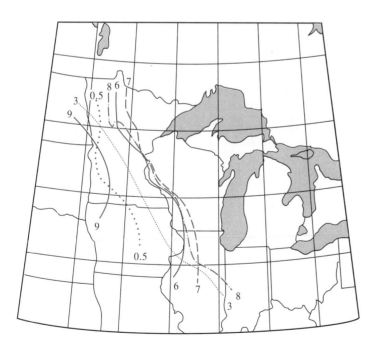

Isochrone curves on a map of the upper Midwest join points having a 20 percent concentration of forb pollen at the same time. The number next to each curve indicates how many thousands of years ago the forb existed at that percentage. The movement of the prairie-forb boundary is apparent.

such three-dimensional representations corresponding to tran-
sects along different latitudes shows how the general nature of the
vegetation changes as we move northward. To examine the
changes over a period of years as we move westward, we would
set up a different series of slices. For points of the original domain
with fixed longitude, we may compare two different three-dimensional representations, or a whole series of them.

We would like to present our data in such a way that we can
stop the progression along three-dimensional graphs for increasing longitude and examine one particular three-dimensional display in greater detail. We would like to be able to view nearby
graphs at the same time, and to achieve some sense of the whole.
The solution is provided by the device of projection, one of the
standard ways of converting configurations in three-dimensional
space into two-dimensional representations on a computer screen
or simply on a sheet of paper. If we project portions of the four-
dimensional display into three-space, we obtain a sort of overlay
effect, as we see two nearby three-dimensional objects overlapping in three-space, with a small shift along an axis, much the
same as the effect of drawing a cube by first drawing the bottom
square and then the top one, shifted slightly in an oblique direction. Studying such two-dimensional representations of configurations in three-space is the necessary preliminary to analyzing
the analogous three-dimensional oblique projections of configurations in a data space of four (or more) variables.

We can slice another way by fixing the time coordinate. We
then have a three-dimensional coordinate system where the horizontal plane gives the region over which the readings are taken,
and the height of the graph above any given point is the abundance of forb pollen at that particular time. The heights at different points form the curved surface of a function graph in three-
dimensional space. As we change the slice in the time direction,
we generate an animated cartoon showing the changing distribution of pollen over hundreds of years or more. We could display
two pollen types simultaneously, perhaps using colors to distinguish the surface of forb pollen from that of the oak or the spruce.
A film or videotape of the changing configuration would give an
excellent presentation of the data from one viewing point. Ideally
we should be able to stop the film at any point and "walk around
the graphs" to determine exactly how the various quantities are
related at a particular time. Some time in the future, we may be
able to display the graphs in holographic motion pictures, so that

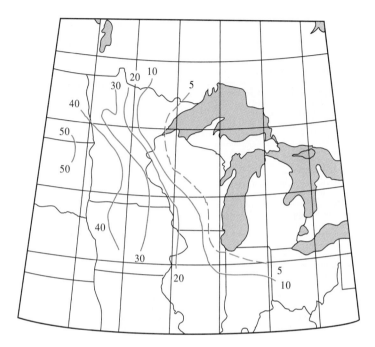

Isopoll curves connect points with the same percentages of forb pollen 6000 years ago. Each curve represents a slice of the four-dimensional data.

each viewer could make his or her own exploratory movements as the film progressed. But even with such a display we would want to have the option of slowing down or stopping the film so that we could investigate a particularly interesting phenomenon at our leisure.

Note that we do not have to take our slices perpendicular to coordinate axes. If we want to examine the data collected along the course of a river valley or even up a mountain ridge, we could slice out a vertical strip over a curve in the two-dimensional region, then flatten it out into the plane for more convenient viewing. The effect is something like a bamboo curtain with three ink markings on every rod. The curtain is curved in one direction so that it fits over the base path, but we might want to flatten it out against a wall so that we could see more clearly how the abundances vary as a function of our position along the transect.

One more kind of slicing is valuable in analyzing data sets in three and four dimensions. Instead of fixing a particular space or time coordinate, we can slice by the coordinate that indicates the abundance of pollen. This means that over the entire domain, for the entire range of time, we might indicate the points at which the abundance of forb pollen is some specific percentage, say 20 percent. Plotting such slices for different abundances yields a series of contour surfaces. On our three-dimensional domain we can attach a number to each point indicating the concentration of pollen at that place and time. We can then connect points of equal concentration, and in general we expect these to fit together in surfaces. If we have 20-percent forb pollen at a particular point, we expect that nearby points will have a 20-percent concentration either at the same time or shortly before or shortly afterward. Thus the data points for nearby sites should fit together to form a small piece of surface over a neighborhood. Of course at a particular site there might be several times when the abundance was exactly 20 percent, so over a neighborhood of that site, the 20-percent contour surface might consist of several pieces. It may be that these pieces come together as we move further from the original point, and the actual arrangement of points in the contour surface might be quite complicated, just as the 200-foot contour line on a landscape might be very involved, if not in the Midwest then certainly in a place like Monument Valley.

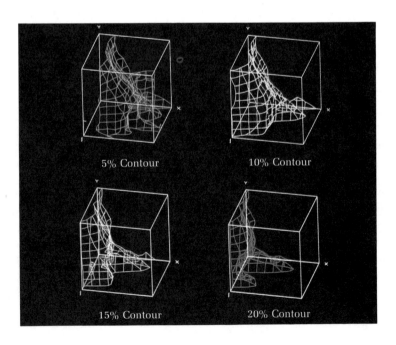

These three-dimensional graphs show four surfaces formed by different concentrations of forb pollen.

There is another problem with this mathematical model. Our precision of measurement is not usually such that we can tell which points have an abundance of exactly 20 percent. At best we can hope for an approximate value, so we may more realistically ask when the abundance falls within a certain tolerance, say between 15 and 25 percent. Instead of a precisely defined surface, we then have a rather indistinct region that contains the surface we are interested in. Often the shape of this region can already give us the information we need to analyze the composition of the flora in a given region over a period of time. There are various averaging techniques which can be used to present these data more clearly.

Once we have a representation of the 20-percent surface, we can investigate it in different ways corresponding to methods used by mathematicians to analyze geometric loci in three-space. Once again the key approaches involve projections and slicing. We can slice by a particular time and see what the 20-percent contour looked like then, and we can try to determine at which points the contour was advancing most rapidly, or where it was receding, questions connected with the gradient of a function of two or three variables. We can "ride the crest" and imagine the progress of 20-percent forb surface as it headed east 8000 years ago.

We can overlay several different data sets to give us a more vivid picture of the interaction of different species. We can look at the 20-percent forb contour in comparison with the contours for 10-percent red oak or 15-percent blue spruce. Or we can color the 20-percent forb surface to indicate the distributions of these other types of pollen, from a light pink to a deep red for the increasing levels of oak pollen, or light to deep blue for the spruce. If we overlay the red and the blue ranges, we obtain a collection of shades of purple, and a glance at a color key could tell us exactly what abundances of red and blue would produce that shade. In this way—by getting a feel for what the data can tell us—we are even more significantly increasing our ability to handle data sets of greater and greater dimensionality. We can construct theories to account for the regularities we perceive, and we can test those hypothetical constructs by further exploration of our data and of similar data sets collected elsewhere. As computer technology goes forward we have new and powerful tools to aid us in this enterprise, and we can look forward to greater and more imaginative insights in the future.

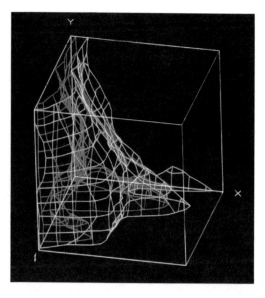

One three-dimensional graph shows surfaces for two different pollen concentrations.

5 REGULAR POLYTOPES and FOLD-OUTS

Like many cultures before and since, the ancient Greeks were fascinated by polygons, planar figures bounded by line segments. Familiar polygons like squares and equilateral triangles and regular hexagons appear in all sorts of planar designs and in architectural constructions. These basic shapes combine with other polygons to form repeating patterns in the plane, and they combine in space to form three-dimensional polyhedral figures such as cubes and pyramids.

Of particular interest are the regular polygons and regular polyhedra, objects with the maximum amount of symmetry possible in their respective spaces. A "regular" polygon or polyhedron looks exactly the same at every vertex, a very strong restriction. In the plane, there are infinitely many regular polygons, each having a different number of sides. But in three-dimensional space, there are only five regular polyhedra. By the middle of the nineteenth century, geometers realized that there could be regular figures in dimensions four and higher, and they wondered how many and of what sorts. The race was on to find the answer to this challenging question, and after a few false starts, several mathematicians each claimed to be the first to solve the problem. The resolution of the dispute surprised everyone involved, as we will see by the end of this chapter.

The Mazar-i-Sharif Hazrat Ali Shrine in Afghanistan exhibits many different polygonal patterns.

A computer graphics re-creation of the first 2 of
16 versions of a polygonal progression designed
by the Swiss artist Max Bill in 1938. The equal-
sided polygons progress outward from the center,
from a three-sided polygon at the heart to an eight-
sided polygon on the outer edge.

The Greek Geometry Game

The Greeks set up a particularly challenging set of rules when
they formalized their notions of geometry. They restricted the set
of tools that could be used for constructing figures to two instruments: a straightedge (with no markings on it) and a compass. The
straightedge enabled geometers to draw a line through any two
points, and the compass enabled them to draw a circle that had a
given point as center and that passed through another given point.

For the Greeks, as well as for today's geometers, a polygon
was a plane figure bounded by a finite number of line segments,
and a regular polygon had all side lengths equal and all angles
equal. Furthermore, the Greeks required that a polygon not intersect itself and that all its diagonals should lie inside the figure. In
order to construct a regular polygon having a certain number of
edges, it is necessary to subdivide the circumference of a circle
into a number of arcs of equal length. In certain cases, this is easy
to do using only a straightedge and compass. In other cases, this
construction is somewhat more involved, and in still others, it is
completely impossible.

The first theorem in Euclid's *Elements*, the most famous textbook of all time, provides a method for constructing a regular
triangle. We start with a red segment for one side of the triangle
and use a compass to describe two orange circles, each with one
endpoint of the segment as a center and the other endpoint on the
circumference. These two circles meet at two points, and each of

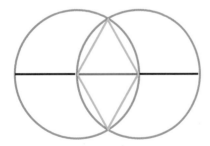

The first theorem in Euclid's *Elements*, the
construction of the equilateral triangle.

these points is the third vertex of a green equilateral triangle having the given segment as base.

We can then extend the red base to intersect the left circle at another point, and from this point as center, we describe a third orange circle of the same radius, meeting the middle circle in two additional points. In this way we obtain six equally spaced points on a circle, forming the vertices of a green regular hexagon. By taking every second vertex as we go around, we divide the circumference into three equal parts, giving the vertices of a blue regular triangle inscribed in the middle circle.

Once we have a regular polygon of a certain number of sides, it is easy to construct one with twice the number of sides just by bisecting each arc using the straightedge and compass. We draw two full circles, orange and yellow, each having an endpoint of the red arc as its center and passing through the original center. These circles meet at two points, the original center and a new point. If we connect these points with the straightedge, the new green line intersects the original red arc in its midpoint.

By bisecting the six sides of the regular hexagon we obtain 12 points equally spaced on the circle. These points are the vertices

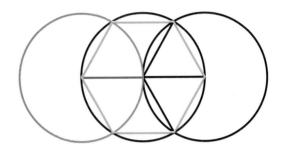

Construction of the regular hexagon inscribed in a circle.

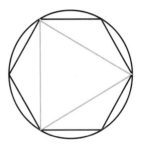

A triangle inscribed in a hexagon.

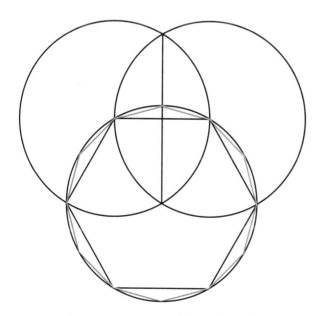

Using bisection to construct the regular dodecagon.

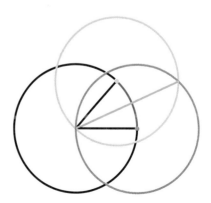

Euclid's argument for bisecting an angle.

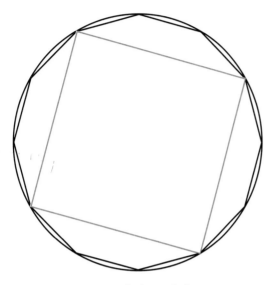

A square inscribed in a dodecagon.

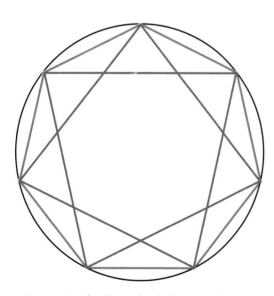

Connecting the diagonals of a heptagon forms a heptagonal star.

of a blue regular dodecagon. Similarly, we obtain a regular polygon with 24 sides, or 48, or an arbitrarily large number, simply by dividing the sides of a polygon in half over and over again. Thus the number of regular polygons that can be constructed just using a straightedge and a compass is unlimited. We can state this more emphatically: there are infinitely many regular polygons in the plane.

In this sentence the words "there are" say something extremely important about the nature of mathematical objects. No one has ever seen a perfect regular triangle, although we have a procedure for generating a representation of one that is as close to perfect as we wish. The concept of "regular triangle" or "regular polygon with four or five or seven sides" makes sense. Such objects have an abstract existence, whether or not anyone ever tried to draw one. We do not know if anyone, using any method whatsoever, has ever constructed a regular polygon with 3072 sides, but we know it can be done and we know how to do it, by starting with the regular triangle and applying the bisection procedure ten times.

We can easily construct other familiar regular polygons, like the square, once we know how to erect a perpendicular to a line from a point on the line. Or we can find a square inscribed in a given circle by joining every third vertex of a regular dodecagon. Once we have the square, we can use bisection to get an entirely new infinite family of regular polygons with 8 sides, 16 sides, and so on, for any power of two.

These relatively easy methods provided constructions for the regular polygons with three, four, six, and eight sides. What about five, seven, and nine? The Greeks solved the problem for a regular pentagon by means of a clever construction related to the geometric solution of a quadratic equation. But no matter how hard they tried, they were unable to construct the regular heptagon with seven sides or the regular enneagon (or nonagon) with nine sides.

It is important to distinguish between an approximate solution and an exact one. The insignia of the Los Angeles Police Department is a seven-pointed star connecting seven equally spaced points on a circle. It is possible to find the location of such points up to any desired degree of accuracy by trial and error, but it is not possible to find the points exactly, as we did for the triangle, square, and hexagon, using just straightedge and compass. Many people who proposed solutions to this problem, including some philosophers and amateur mathematicians, simply did not understand the difference between an approximation and an exact solution.

It was especially embarrassing for the Greeks that their favorite geometrical methods were unable to provide a construction for the regular enneagon. Greek geometers had already divided the circumference into three equal arcs. Now all they had to do was divide each of these arcs into three equal pieces. Dividing the segment across the arc into three equal parts and extending rays through these points does not work since the middle arc will be larger than the other two.

Even though there is a straightedge-and-compass method for bisecting an arbitrary angle, there is no such method for trisecting an arbitrary angle. In fact there is no possible straightedge-and-compass method for trisecting the particular angle that is one-third of a circumference, or 120 degrees. When a hopeful amateur submits a "solution" to the trisection problem, a flaw in the argument usually shows up when the proposed method is applied to this particular angle. (This does not necessarily dissuade the would-be solver.)

Not until the early part of the nineteenth century did mathematicians develop the algebra that proved the impossibility of trisecting the 120-degree angle. The straightedge-and-compass method for bisecting an angle involves finding intersections of circles, and there is a corresponding algebraic process, which involves taking square roots and solving quadratic equations. The geometric problem of trisecting an angle corresponds to the algebraic problem of solving a cubic equation, and as mathematicians discovered, the straightedge and compass cannot provide the solutions of all such equations. For the 120-degree angle, the argument involves a trigonometric identity that gives the relation between the cosine of an angle and the cosine of one-third of the angle. Finding the cosine of the 40-degree angle is equivalent to solving the cubic equation $8x^3 - 6x + 1 = 0$, and this cubic equation cannot be solved just by taking repeated square roots.

Does that mean that the nine-sided regular polygon does not exist? Not at all. It exists just as much as a triangle or a square or a circle exists, but it cannot be constructed with straightedge and compass, so we cannot expect to encounter it among the objects of Euclid's plane geometry. Existence is not the same as constructibility. There exist infinitely many different regular polygons, one for each number greater than two, even though we can't construct them all using the classical Greek methods.

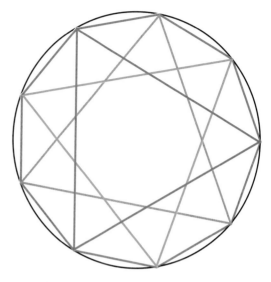

Three triangles inscribed in an enneagon.

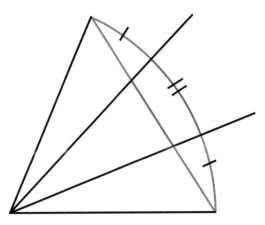

Trisecting a chord does not produce a trisection of an angle.

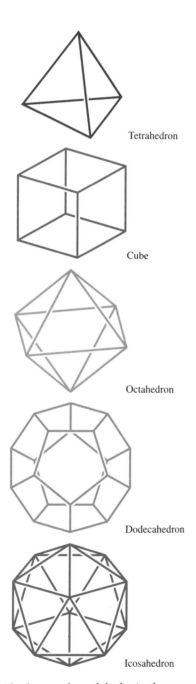

Tetrahedron

Cube

Octahedron

Dodecahedron

Icosahedron

The five regular polyhedra in three-space.

The Search for Regular Polyhedra

In contrast with the unlimited number of regular figures in the plane, the number of regular objects in three-space is small. The Greek geometers not only found them all, they also proved that there weren't any more to be found.

Long before Greek mathematicians formalized the axioms for solid geometry, people were familiar with several regular polyhedra, in particular the cube, the tetrahedron (the Greek term for a figure with four faces), and the octahedron (a figure with eight faces formed by putting together two square-based pyramids with equilateral sides). If we wished to follow the Greek usage completely, we should call the cube a hexahedron, but we will continue to use the familiar Latin expression. The cube has three squares at each vertex, the tetrahedron has three equilateral triangles at each vertex, and the octahedron has four equilateral triangles at each vertex. All the faces of a regular polyhedron must be regular polygons, and there must be the same number of faces meeting at each vertex. Do any other solid figures satisfy these requirements?

By the time Euclid wrote his textbook, two more regular polyhedra had been discovered. In the thirteenth and final book of his *Elements*, Euclid included a proof that there could not be more than five regular polyhedra. It is worth our while to consider the proof carefully because it contains an idea that can assist us when we ask how many regular figures exist in higher dimensions.

Euclid first observed that for a regular polyhedron the angles that come together at a given vertex add up to less than 360 degrees. For example, the three squares at a vertex of a cube have an angle sum at the vertex equal to three right angles, or 270 degrees. Next he noted that in order to make a polyhedron, at least three faces must meet at each vertex. If we want to make a regular polyhedron with triangular faces, we then have only three possibilities: three, four, or five triangles at each vertex. Six triangles will already fill the area around a point, leaving no gap that would allow us to fold the object up in three-space.

Three triangles meeting at a vertex fold into a triangular pyramid, and adding one more face gives the tetrahedron. Four triangles meeting at a vertex fold into a pyramid with a square base, and putting two of these pyramids together along a common square face gives the octahedron. Five triangles meeting at a ver-

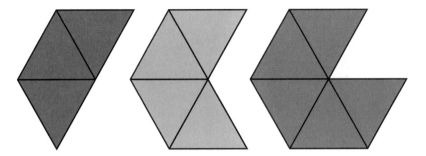

Possible arrangements of triangles around a point in the plane.

tex fold into a pentagonal pyramid with a regular pentagon as base and five equilateral triangles as sides. To construct a regular polyhedron that includes this figure, we begin with a strip of 10 equilateral triangles alternately pointing up and down and two regular pentagons with the same side length. By fitting these pieces together, we construct a pentagonal antiprism: one regular pentagon is the base; the other regular pentagon, slightly rotated, is the top; and the strip of triangles is sandwiched between such that each edge of a pentagon is an edge of an equilateral triangle having its third vertex on the other pentagon. We then erect pentagonal pyramids on the top and bottom pentagons to obtain a figure with 20 equilateral triangles, called a regular icosahedron.

This completes the list of regular polyhedra with triangular faces, but what about other regular polygons? It is possible to fit three squares around a point, but four already fill the area around a point, so the cube is the only regular polyhedron with square

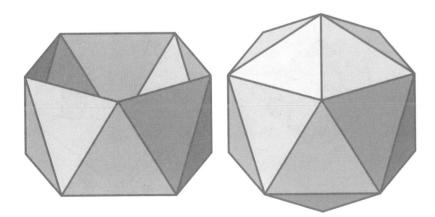

A pentagonal antiprism and the completed icosahedron.

faces. We will not find any regular polyhedron with hexagonal faces since three hexagons already fill the area around a point. Three polygons with more than six sides will more than fill the space around a point.

The only candidate left is the regular pentagon. Since the angles of a pentagon are less than those of a hexagon, three pentagons will fit around a point in the plane with room left over. Since the angles of a regular pentagon are greater than those of a square, we cannot fit four regular pentagons around a point in the plane. This leaves open the possibility of a fifth regular polyhedron having three regular pentagons around every vertex. Well before the time of Euclid, Greek geometers had found this fifth regular polyhedron, a regular dodecahedron having 12 pentagonal faces.

Duals of Regular Polyhedra

One of the most illuminating ways of constructing a regular dodecahedron is by appealing to the principle of duality. A cube and an octahedron, for example, are very closely related. If we choose the centers of the six square faces of a cube, these are the vertices of an octahedron. We say that the octahedron is the dual of the cube. Conversely, the centers of the eight triangular faces of an octahedron are the vertices of a cube, so the cube is the dual of the octahedron.

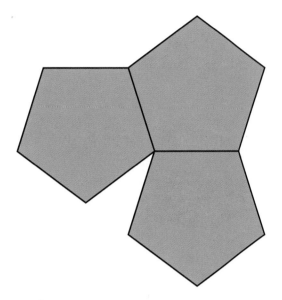

Three pentagons around a point in the plane.

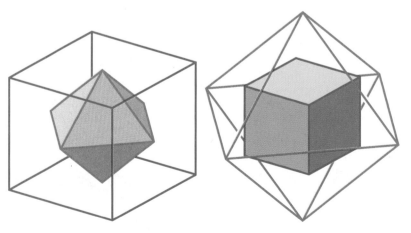

The octahedron is dual to the cube.   The cube is dual to the octahedron.

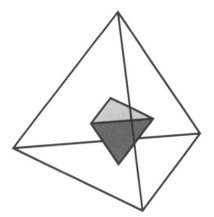

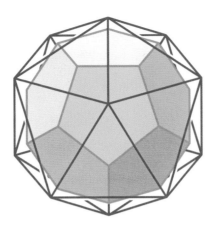

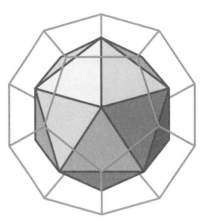

The self-dual tetrahedron.

The dodecahedron is dual to the icosahedron.

The icosahedron is dual to the dodecahedron.

What happens if we construct duals of other regular polyhedra? For a tetrahedron, the centers of the four triangular faces form another tetrahedron, so the tetrahedon is self-dual. More thought is required to find the figure having as vertices the centers of the 20 triangles of the icosahedron. Around each vertex of the icosahedron there are five triangles, and the centers of these five triangles when connected form a regular pentagon. The icosahedron has 12 vertices, so we obtain a regular arrangement of 12 regular pentagons, three at each vertex. This is the fifth regular polyhedron predicted by our argument, the regular dodecahedron.

The regular dodecahedron has 20 vertices, with three pentagons at each vertex. The centers of the pentagons will then give 20 equilateral triangles, forming a regular icosahedron. Thus the five regular polyhedra fall into three groups: two dual pairs and one polyhedron that is dual to itself.

The Search for Regular Polytopes

A group of Flatlanders could easily follow Euclid's argument for determining the number of regular polyhedra. They could understand the theorem that there are at most five ways to fit copies of the same regular polygon around a point in their flat space. They would not be able to imagine what it would be like to fold such a configuration up into three-space, but they could still appreciate that there are at most five regular polyhedra in three-space.

Once mathematicians realized they could think about geometry in higher dimensions, they began to look for analogues of polygons and polyhedra. Just as polygons are bounded by segments and polyhedra are bounded by regular polygonal figures, the analogous objects in four dimensions would be bounded by regular polyhedra. Such objects in higher dimensions came to be known as polytopes.

The discovery of all five regular polyhedra in three-dimensional space was well known and well appreciated. It was natural

Illustrations of William Stringham's approach to the search for regular polytopes, from the third volume of the *American Journal of Mathematics*, 1880.

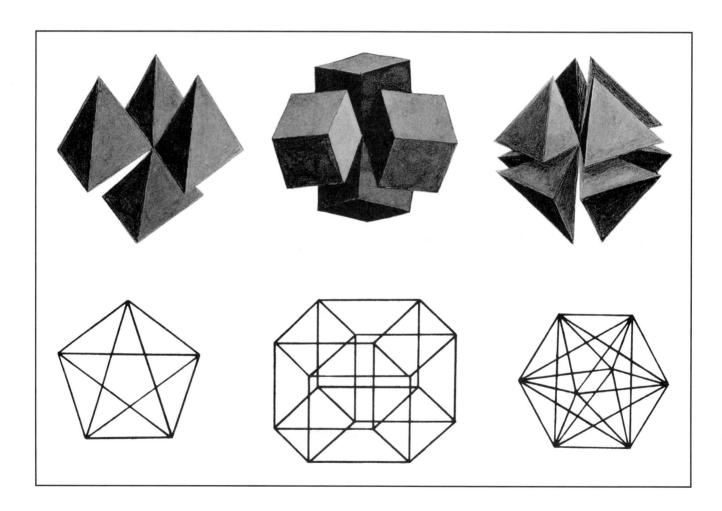

to try to find the analogous result in four-dimensional space, and the "search for the regular polytopes" was on. In the 1880s, the decade in which Abbott wrote *Flatland*, there was a veritable polytope rush among mathematicians in the United States, Scandinavia, and Germany. At least one reputable mathematician published an incorrect list, and intense argument erupted about who was the first to find all of the regular polytopes. The American contender, William Stringham, analyzed the possible configurations of regular polyhedra around a point in three-space and produced a set of pictures, which were printed in his article in the *American Journal of Mathematics*. But there were a large number of cases to consider and his argument was incomplete. He was not completely convinced he had found all of the regular polytopes. Fortunately there soon appeared a simpler and more compelling argument.

To understand Stringham's approach, consider the regular figure that we already know very well, the hypercube. It has 16 vertices, with four edges at each vertex. Any three of these edges determine an ordinary cube, so there are four cubes at each vertex. Just as we can think of the part of a cube near a vertex as three squares in the plane, with instructions for attaching two edges together in three-space, we may think of the part of a hypercube near a vertex as four cubes in three-space, with instructions for attaching pairs of square faces together. And just as the Flatlanders cannot actually put the three squares together, since they do not have access to our third dimension, we cannot put the four cubes together to form part of the hypercube in four-space, but we can still appreciate the problem.

The figure in the margin of this page suggests an approach that simplifies the problem of determining possible regular polytopes in four-space. Instead of examining the many possible ways in which a collection of polyhedra can fit around a vertex, rather consider the number that can fit around an edge. When the sum of the angles of the polyhedra around an edge does not fill the space around the edge, there is room to fold the figure up into the fourth dimension.

The problem of determining how many regular polyhedra fit around an edge turns out to be relatively easy to investigate by experiment. It is clear that one can get three cubes around an edge in three-space, and that four will already fill the region. Therefore, there can be at most one regular polytope with cubical faces, namely the hypercube.

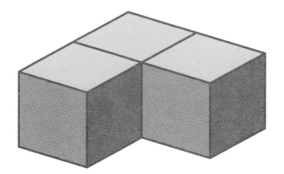

Three cubes around an edge in three-space.

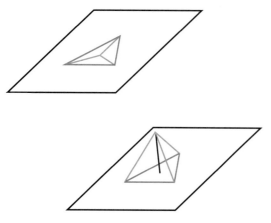

Construction of a three-simplex by lifting the center of a two-simplex.

The Four-Simplex

What about polytopes made out of tetrahedra? In the previous chapter, we discussed a method of building one such object, the four-dimensional simplex, or pentatope. This object is the simplest polytope in four-space, analogous to the triangle in the plane and the tetrahedron in three-space. To construct a four-simplex, start with a segment in the plane, and draw the line perpendicular to it through its midpoint. Any point on this line is equidistant from the two endpoints. If we go out far enough, the distance to the endpoints is equal to the length of the segment and we have a regular triangle, also called a two-simplex.

Now draw the line in space perpendicular to the triangle through its center. Although the Flatlanders cannot appreciate this construction, we know that any point on this line is equidistant from the vertices of the triangle, so if we go out just far enough, we find a vertex with distance equal to the side lengths of the triangle. This produces three new equilateral triangles congruent to the original triangle, and we have constructed a regular tetrahedron, or three-simplex.

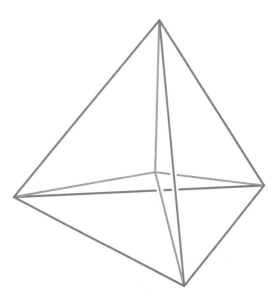

Symmetric projection of a four-simplex into three-space. Displacing the center point perpendicularly into four-space makes all 10 edges equal in length.

In a similar way, we may consider a line in four-dimensional space perpendicular to the tetrahedron through its midpoint. We in three-space cannot see this line, but we know that any point on it is equidistant from all vertices of the tetrahedron. If we go out along this line just far enough, we find a point with distance equal to the lengths of the sides of the original tetrahedron. In this way we obtain four new tetrahedra, each congruent to the original tetrahedron, forming a regular four-simplex in four-dimensional space.

We do not have physical access to that fourth direction, so we cannot actually construct the four-simplex. Nevertheless, using the techniques of the previous chapter, we can still draw pictures of its shadows in the plane or in three-space. When we draw a tetrahedron in the plane, we choose four points and connect all possible pairs. We get two different sorts of views, depending on whether or not one vertex is situated inside the triangle formed by the other three. Similarly there are two different sorts of projections of a pentatope into three-space, depending on whether or not one vertex is contained inside the tetrahedron formed by the other four. The first example on this page has four vertices on the outside and one inside, six edges outside and four inside, and four

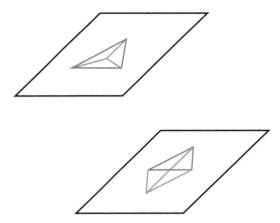

Two projections of the three-simplex into two-space.

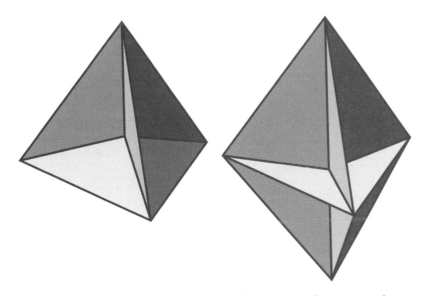

Two projections of the four-simplex into three-space, with inner triangles filled in.

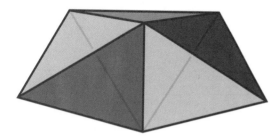

Three tetrahedra around an edge in three-space.

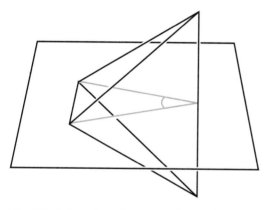

The dihedral angle at the center of one edge of a three-simplex.

Four tetrahedra around an edge in three-space.

triangles outside and six inside. In the second example, all vertices are outside, all edges but one are outside, and there are six triangles outside and four inside. There is a major difference between the two representations. In the first it is possible to put in all ten triangles without having any one intersect any other, while in the second the three vertical triangles and their common edge all intersect the horizontal triangle.

Either projection of the four-simplex shows three tetrahedra at each edge, just as there are three cubes at each edge of a hypercube. Analogous to the configuration of three cubes around a line, we may arrange three tetrahedra around an edge in three-space, with room to spare.

To see how many regular polyhedra can be placed around an edge in three-space, we sum the dihedral angles of each polyhedron, the angles between the two planar faces of the polyhedron that come together at the edge. In the case of the tetrahedron, we can think of the dihedral angle as a planar angle by holding the tetrahedron with one edge vertical and the opposite edge horizontal, and then slicing through the middle to get an isosceles triangle. The angle of this triangle that lies on the vertical edge will be the dihedral angle at the edge. It is larger than the angle of an equilateral triangle, so six tetrahedra cannot fit around an edge.

The Hypercube Dual or Sixteen-Cell

The dihedral angle of a tetrahedron is less than a right angle, so we can easily fit four tetrahedra around an edge in three-space, with room to fold up into four-space. There is a regular polytope that has this configuration of four tetrahedra at each edge, and to describe its construction we rely on the duality principle. To construct the dual of a polytope, we choose the vertex in the center of each three-dimensional face, and we connect vertices from faces that share a common boundary polygon. The vertices coming from three-dimensional faces sharing a vertex determine a three-dimensional polyhedron, the dual cell of the vertex. The collection of dual cells forms the dual polytope of the regular polytope.

Like the three-simplex or tetrahedron, the four-simplex is self-dual. To describe the dual of the hypercube, we choose a point in the center of each of the eight cubical faces. Around each of the 16 vertices there are four cubes, each set of four contributing a tetrahedron to the dual object. We thus obtain a new regular polytope, the 16-cell, dual to the hypercube, and analogous to the octahedron, which is dual to the cube in three-space. This polytope has four tetrahedra around each edge, and it is the third of the regular polytopes in four-space.

Polytopes in Five or More Dimensions

The above constructions are in no way special for four-space. In every dimension there is a self-dual simplex, with $n + 1$ vertices when the dimension is n. Also in every dimension is an analogue of the cube. In n-dimensional space, the n-cube has 2^n vertices, and it has $2n$ highest-dimensional faces of dimension $n - 1$. There will always be a third regular polytope in n-space, the dual polytope to the n-cube, with $2n$ vertices and 2^n highest-dimensional faces, which are simplices of dimension $n - 1$. These constructions will become clearer when we introduce coordinates in Chapter 8.

As it happens, for n larger than four, this is all we get. In n-space, there are exactly three regular n-dimensional polytopes, the n-simplex, the n-cube, and the n-dimensional cube-dual. There are no further regular polytopes.

The Regular 600-Cell and Its Dual

But in four-space there is a surprise. We have found three tetrahedra around an edge, and four, and we know that there cannot be six. What about five? It turns out that the dihedral angle of a tetrahedron is just small enough that we can fit five tetrahedra around an edge with a very small amount of room to spare, which allows for folding into four-space.

This configuration of five tetrahedra around each edge is found in a four-dimensional polytope having 600 tetrahedra and therefore going by the name of a 600-cell. We can describe the part of this polytope near a vertex by using the same construction we used to get from a pentagon to the pentagonal pyramid at each vertex of a regular icosahedron. For the icosahedron, we start with

Projection of the 16-cell into the plane, showing 2 of the 16 three-simplexes. The others are obtained by rotating the figure by multiples of 45 degrees.

Five tetrahedra around an edge in three-space.

Projection of the 600-cell into three-space, from
Regular Complex Polytopes by H. S. M. Coxeter.

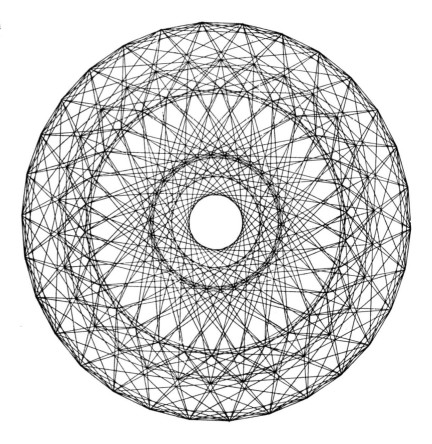

a regular pentagon in the plane and choose a point above its center
so that the distance to each of the five vertices equals the length of
the original edges. For the 600-cell, we begin with an icosahedron
in three-space and choose a point "above" its center in four-space
so that the distance to each of the 12 vertices equals the length of
the original edges. Drawing the edges from this point to the original vertices forms 20 tetrahedra coming together at the vertex of
the 600-cell. Once again, we cannot actually carry out this construction without access to a fourth direction perpendicular to our
space, but the procedure is clear.

   Every regular polytope should have a dual, and this dual will
be another regular polytope. Since each vertex of the 600-cell is
surrounded by 20 tetrahedra, each cell of the dual polytope will

have 20 vertices. Thus the dual cells will be dodecahedra. It may seem surprising that a regular polytope can be built out of dodecahedra since three regular dodecahedra would have to fit around an edge in three-dimensional space. Yet indeed they do fit. The dihedral angle of a dodecahedron turns out to be slightly less than one-third of a complete rotation, so three of them fit around an edge with a small amount of room left over. The dual of the 600-cell turns out to have 120 regular dodecahedra, leading to its name, the 120-cell.

We had five regular polyhedra in three-space, and the above analysis has already produced five regular polytopes in four-space. We don't get any new examples by using icosahedra since the dihedral angle at an edge of a regular icosahedron is larger than one-third of a full rotation. But still remaining is another possible building block, the octahedron, and it provides one more surprise.

The Self-Dual 24-Cell

The dihedral angle of the octahedron is larger than that of the cube and smaller than that of the dodecahedron, and so we can fit three octahedra around an edge but not four. It follows that there is a possible regular polytope in four-space with octahedral faces. Such an object does exist. Having 24 octahedral cells, it is called the 24-cell (see illustration on the following page). Around each vertex of such an object there will be six octahedra. Therefore the cells of the dual polytope each have six vertices, and these dual cells too are regular octahedra. The dual of the 24-cell is another 24-cell, so this polytope is self-dual. The surprise then is that there are more regular objects in four-space than there are in three-space, even though in higher dimensions we only find the three basic types.

According to Abbott, contemplation of higher dimensions leads to humility. That certainly was true for the many mathematicians in the 1880s who vied for the honor of being the first to find all of the four-dimensional regular polytopes. We have proven that there can be at most six regular polytopes in four-space, and so did they. Which of them deserved the credit for the discovery? The entire competition turned out to be meaningless when it became known that the result had been proven more than

Projection of the 120-cell into three-space created by Paul R. Donchian, part of the collection of the Franklin Institute in Philadelphia.

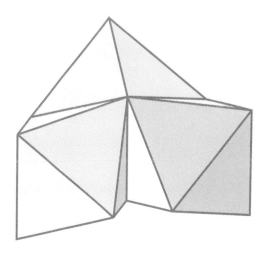

Three octahedra around an edge in three-space.

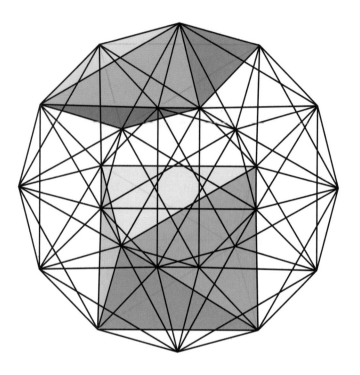

Projection of the 24-cell into three-space.
Highlighted are the differently shaped projections
of two octahedra.

30 years earlier by the Swiss mathematician Ludwig Schläfli, in
a long work on higher-dimensional geometry which contained not
one picture!

Fold-Out Patterns in Different Dimensions

Most of our argument so far has been in terms of the local structure of a polyhedron or a polytope—that is, the faces in the neighborhood of a point or an edge. In order to gain a better appreciation of the global structure, the shape of the object as a whole, we
can use the device of fold-out patterns.

To illustrate this device, we go back to lower dimensions. It
would be possible to give the King of Lineland a kit for building a
square, just four segments of equal length with instructions telling
which endpoints should be attached to each other. The King
could begin the assembly by making three of the attachments, but
he would be left with just a hinged rod four times the length of the

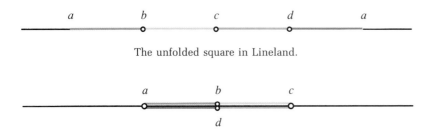

The unfolded square in Lineland.

Overlapping construction of a collapsed square.

sides of the square. He could not make the fourth connection without having the sides overlap. If it were possible for two sides to occupy the same space on the line, the King could just place one hinged rod with two sides on top of a similar figure and connect their endpoints. But he still could not construct a true square in the line since this "collapsed" quadrilateral has two different kinds of angles, two of them straight angles and two of them zero angles. Of course these are the only kinds of angles available in Lineland. We have to go to the plane to obtain a square, having all edges the same length and all angles equal.

We encounter similar challenges when we move from the plane into three-space. We can design a prefabricated polyhedral structure in the plane by laying out the polygons and indicating which edges are to be attached to which. A preparatory crew in Flatland could begin the assembly, but they could not complete the project.

A good example is the fold-out pattern for a cube. We place down one of the square faces and attach the four adjacent squares to obtain a cross-shaped pattern. We then indicate which sides of these squares are to be attached to which, giving the fold-out pattern for an open box. There are exactly four free edges not yet attached to anything else, and these are ready for the last square, which we place at the bottom of the cross pattern. An engineer in Flatland could supervise the accurate construction of the individual pieces, but when it came time to assemble the object, it would be necessary to create unstraight dihedral angles, and all but the original square would simply disappear as the other squares folded into three-space. If some faraway light caused the folding cube to cast a shadow, each of the four side squares would be replaced by a shadow rectangle, which would shrink back to a side of the original square.

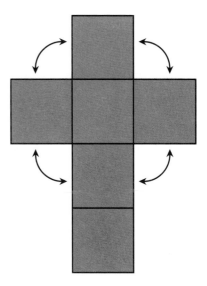

The unfolded cube in the plane.

A hypercube kite, created by José Yturralde in Valencia.

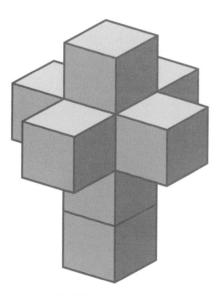

The unfolded hypercube in three-space.

The analogous figure in three-dimensional space is an unfolded hypercube. Our three-space engineers could manufacture the eight cubical faces of the hypercube, and start the construction by arranging a cube surrounded by six cubes. After we see how neighboring square faces match up, there will remain six unattached squares. These are ready to receive the eighth cube, which we place at the bottom of the figure. Unfortunately no one from our space can fully appreciate the process that puts this hypercube together. Analogous to the previous case, if some faraway

The Crucifixion, subtitled *Corpus Hypercubicus,* painted by Salvador Dalí in 1954.

light in four-dimensional space caused the folding cubes to cast shadows in our space, we could watch as each of the six cubes became a shadowy rectangular box that gradually flattens out to one of the faces of the original cube.

In 1954 Salvador Dalí used the unfolded hypercube as the main symbol in his painting *Corpus Hypercubicus,* representing a Christ figure suspended in front of this unfolded cross-shape from the fourth dimension. In 1976 Dalí contacted us at Brown University to discuss some of the mathematics in a project he was work-

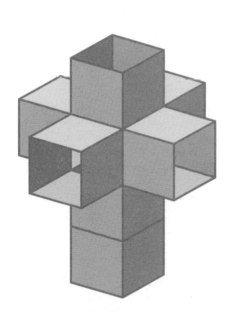

The folding hypercube.

Construction of the folding hypercube.

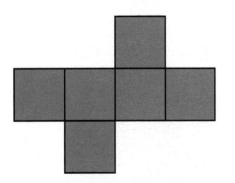

This strip of four squares can be folded to make a square prism, with two additional squares for the top and bottom.

ing on in stereoscopic oil painting, and he very much liked our folding and rotating hypercube model. A copy of this model is on exhibit in the Salvador Dalí Museum in Figueras, Spain.

To construct this model, begin with six square cylinders, each composed of four squares. Attach each of these cylinders to the edges of a cube, and attach a seventh cylinder to the boundary of the bottom piece. The resulting object will flatten into a plane and rotate freely as a sort of "universal joint" when we hold onto two opposite cubes. This model is the primary subject of Robert Heinlein's story, ". . . and He Built a Crooked House," the tale of an architect who builds a home in the shape of an unfolded hypercube. The house suddenly folds up into the fourth dimension, taking with it the shocked owners who have to figure out what happened and how to get back to their own dimension.

The fold-out patterns for regular polyhedra reveal that their polygonal faces seem to arise in strips. We can think of a cube as a strip of four squares fitting together to form a polyhedral cylinder, capped by the two other squares. A strip of four triangles folds together to form a tetrahedron. Six regular triangles form a polyhedral cylinder having two triangular rims in parallel planes. Fill in two triangles and we have an octahedron. The same construction applied to a strip of 10 regular triangles yields a polyhedron with two pentagonal faces in parallel planes. This object is the pentagonal antiprism we used to construct the regular icosahedron. We may carry this construction out in six different ways, describing the icosahedron as a pattern of woven strips. Finally, a strip of 10 pentagons capped by two more pentagons forms the dodecahedron.

To carry this idea into the next dimension, rather than start with a strip of polygons with extra polygonal faces attached, we start instead with a stack of polyhedra with extra polyhedra attached. In this way we can obtain the fold-out models of all of the regular polytopes. An example of this construction for the 24-cell is included in the author's paper in the volume *Shaping Space*. The Canadian geometer H. S. M. Coxeter gives a similar description for the 120-cell and the 600-cell in his book *Regular Polytopes*.

The search for regular polytopes stretches over more than a century, and discoveries are still being made. Some of the properties of these polytopes were identified fairly early, but it has taken a long time to gain a full appreciation of the beauty of these objects. Only with the interactive computer are we able to turn these objects around in front of us, experiencing the same fascination that one of our remote mathematical ancestors felt when he or she first turned over and over again a die in the shape of an octahedron or a dodecahedron. The fascination will continue on into the future as we begin to explore a myriad of intriguing shapes, all beginning with the search for the regular polytopes.

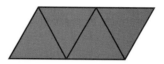

A strip of four triangles can be folded to make a tetrahedron.

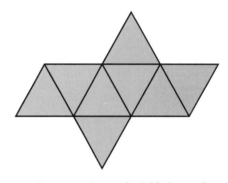

A strip of six triangles can be folded to make a triangular antiprism, with two additional triangles for the top and bottom of an octahedron.

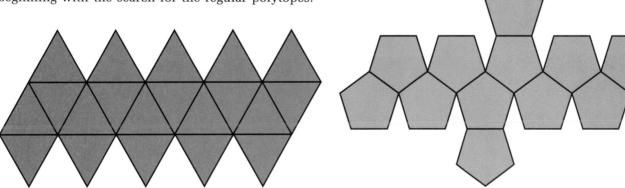

A strip of 10 triangles can be folded to make a pentagonal antiprism, with 10 additional triangles for the top and bottom caps.

A strip of 10 pentagons can be folded to make all but two pentagons of the dodecahedron.

6 PERSPECTIVE and ANIMATION

Salvador Dalí once designed a horse 30 kilometers long. His original plan had not been so grandiose—at first the statue was to have a length of only about a hundred meters. In order to see it the right way, you would come through a gateway and look up to behold the realistic figure of a large horse on a ramp looking down at you. Closest to you would rise the head with its flaring nostrils, then farther back the perfectly proportioned powerful shoulders, and farthest away a large rump. A very realistic looking horse, you would think, until you explored further. In actuality the shoulders would sit several meters away on a high structure, and the rump would rise very high on a several-story construction at the other end of the exhibit area. The rest of the horse would be stretched in between these three parts. Only from one viewing point would the statue appear realistic. From any other point, the horse would look greatly distorted (see Dalí's sketches on the next page).

Dalí did not find it difficult to calculate how large to make the various elements, and just how far back from the observer to place them so that the appearance would be just right, at first glance anyway. It was an exercise in a three-dimensional *trompe l'oeil*, similar to the famous pictures of domes painted on flat ceilings, which deceive viewers for that instant when they first come into a

Parallel lines in the gardens at the Taj Mahal converge to a single vanishing point.

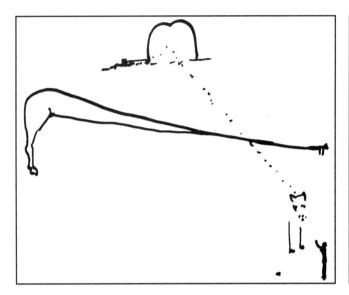

Sketches by Salvador Dalí show the hundred-meter horse in extreme perspective.

Salvador Dalí and the author in 1976.

gallery. Such images look absolutely three-dimensional at first, but they quickly lose their realism when some exploratory motion shows that the image does not behave the way a three-dimensional object should.

Dalí was not satisfied with a horse the length of a football field. His next design expanded the shoulders of the horse and moved them to the top of a tall building at some distance from the head. The rump was to be placed on a conveniently shaped mountaintop 30 kilometers away. The basic mathematical design of the sculpture remained the same—only the scale was different.

The final project went even further—the shoulders were to sit on the mountaintop and the role of the rump was to be played by the moon. Of course only at very rare intervals would the moon come over the mountain at just the right spot to complete a realistic picture of a horse. To view the sculpture, an observer would have to watch not only at the right place on the earth, but at the right time.

The sculpture will not be built. But with computer graphics, we can see exactly what the horse would look like if it ever were built. That appealed to Dalí.

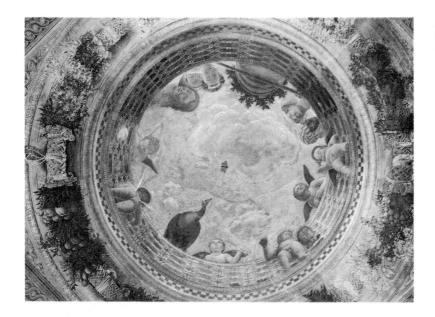

A trompe l'oeil painting of a dome strikes the eye as the real structure until it is viewed from a different angle.

Viewing in Perspective

Dalí's design relied on his mastery of perspective, one of the most useful means of interpreting visual imagery. As we stand in a doorway and look into a room, the way we see the interior furnishings depends very much on our specific viewpoint. If someone wanted to trick us, he could photograph the interior from exactly the same viewpoint and mount the life-size image on a flat screen just inside the room. If we came to the doorway and looked in again, we would see precisely what we had seen before—we would receive exactly the same visual information. How could we tell the difference between the picture and the reality?

The most natural way to tell the difference is to change the viewpoint. Coming a bit closer or moving to the side, you would see how the image shifted. A flat picture will shift quite a bit differently from a real three-dimensional room. In the real room, a square picture frame will begin to look like a trapezoid as we change our position, but in a two-dimensional picture, it will not undergo any such distortion.

Perspective Twist by Lana Posner shows true perspective in each portion of the figure, but an impossible configuration overall.

We can imagine A Square doing the same thing, walking around in his space to determine whether some image he sees is a real view of a house, or merely a representation of the interior of a house painted on a one-dimensional screen. The trompe l'oeil effect should be strong in every dimension, as we can all be fooled by the illusions appropriate to our own space. If we can learn to deal with the illusions of perspective, we can use them to good advantage to help us understand complicated three-dimensional structures, and eventually to visualize objects from the fourth dimension.

Up to this time all of our images of the cube and hypercube have been shadows cast by parallel rays of light. The images of parallel lines appeared as parallel lines (or points), and parallel segments of the same length had images that were segments of the same length. We know, however, that this is not what we actually see when we take in a large view. Parallel railroad tracks seem to converge to a point on the horizon. The railroad ties may appear to be parallel to the horizon, but they get shorter and closer together as they recede into the distance. The reason why a faraway railroad tie looks smaller is that the rays from its endpoint to the observer's eye form a smaller angle than do the rays of an equal-length tie that is closer to the viewing point. In a perspective drawing, any collection of parallel lines will appear either as par-

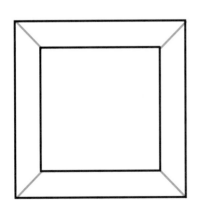

A view of a cube with one vanishing point.

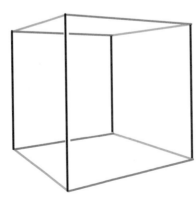

A view of a cube with two vanishing points.

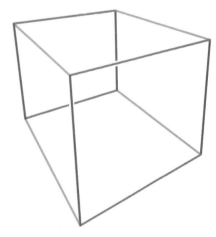

A view of a cube with three vanishing points.

This view from directly above a boxing ring nicely illustrates a "square-within-a-square." Vertical cables converge to the center of a square surrounded by larger squares formed by the images of the ropes. Cleveland Williams is lying flat on the floor, while Muhammad Ali retires to a neutral corner.

allel lines or as lines converging to a vanishing point. A cube has three sets of parallel edges, and the image can have one, two, or three vanishing points.

When we look at a cube, the closer parts will have larger images and the parts farther away will have smaller ones. The front face of a cube viewed head on will be larger than the back face. This "square-within-a-square" is a familiar representation of our view of a cube from in front of one face. Images of vertical and horizontal edges appear as vertical or horizontal segments, but the edges heading away from us appear to converge toward the center. The closest and farthest faces appear to be squares, and the images of the other four faces are trapezoids. From experience we know that the six faces of the cube are squares of the same size and shape, even though they do not appear that way in any single view.

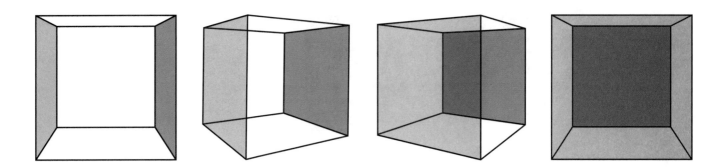

Perspective views of a cube rotating about a
vertical axis.

To get a better picture of the cube, we could look at a whole sequence of views as we walk around the cube, or, equivalently, as we stand still and the cube spins about a vertical axis. As the cube begins to spin, the images of vertical lines remain vertical, but the previously horizontal and parallel segments now have images lying in lines that converge to a point. At this stage, the images of the top and bottom squares are not even trapezoids since no two edges have parallel images.

As the cube continues to rotate, the trapezoidal image of one of the vertical faces appears to flatten out into a vertical segment and then to open again into a trapezoid. The image appears to pass through itself as the inner and outer squares change places. The same phenomenon occurs if we rotate the cube about a horizontal axis.

Perspective always causes some distortion, but we are able to accommodate the distortions by relying subconsciously on our experiences of viewing objects. When we see a rotating cube, we think about a cube, not the varying sequence of squares, trapezoids, and more complicated four-sided figures. The more aware we become of the way we visualize shapes in three-space, the better we can apply the principles of perspective to help us visualize shapes in spaces of dimensions four and higher.

Perspective Views of the Hypercube

Analogous to the perspective views of a cube, we can imagine a sequence of perspective views of a hypercube in three-dimensional space. Just as a cube appears to be a square within a square when viewed from directly in front, the head-on view of the hypercube will appear to be a cube within a cube. The closest part of the hypercube will appear as a large cube, and the part farthest away will appear to be a smaller cube inside the larger one. In the three-dimensional case, the images of four edges of the cube join vertices of the outer square to corresponding vertices of the inner square to form four trapezoids. In the four-dimensional case, the images of eight edges of the hypercube will join vertices of the outer cube to corresponding vertices of the inner cube, thus forming six incomplete pyramids.

This central projection is one of the most popular representations of the hypercube. It is described in Madeleine L'Engle's novel *A Wrinkle in Time* and in Robert Heinlein's classic short story ". . . and He Built a Crooked House." Some writers refer to this central projection by the name *tesseract*, a term apparently going back to a contemporary of Abbott, C. H. Hinton, who wrote an article "What Is the Fourth Dimension?" in 1880 and his own two-dimensional allegory, *An Episode of Flatland*, the same year that Abbott wrote *Flatland*. The sculptor Attilio Pierelli used this projection as the basis of his stainless steel "Ipercubo."

Difficult as it is to imagine the perspective views of a slightly rotated three-dimensional cube, it is even more difficult to imagine the views of a hypercube as it spins in four-dimensional space. Fortunately the computer, which provided the parallel projections of the hypercube studied in Chapter 4, can also produce perspective views of the hypercube from any viewpoint we choose. The method by which it achieves this result is called central projection.

If we want to make a picture re-creating the perspective view of a cubical framework viewed from the top, we can shine a bright light from our viewing point and capture the shadow on a photographic plate below. That shadow gives an accurate record of exactly what would be seen from the viewpoint at which the picture was taken, and if we stand in that spot and look at the shadow image, we would receive the same impression as we would get from looking at the actual object.

It is not difficult to instruct a computer to make such images on a graphics screen. Once we choose a viewing point and a projection plane, the machine can determine the position of the image of a vertex by figuring out where the ray from the viewpoint through the vertex strikes the plane. Almost the same mathematics that creates a central projection of a cube on the plane can be used to create a central projection of a hypercube in three-space,

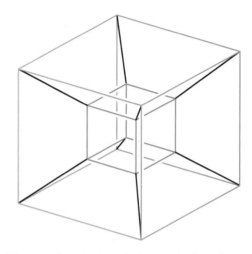

The central projection of a hypercube from four-space to three-space appears as a cube within a cube.

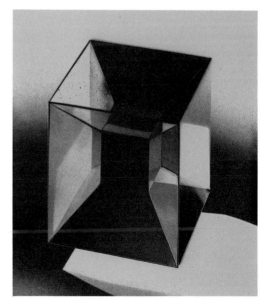

A stainless steel sculpture of a central projection of the hypercube by Attilio Pierelli.

and then to show the projection on the computer screen. Before we discuss the problems of visualizing the three-dimensional shadows of four-dimensional objects, we will consider the application of central projection to other figures in three-dimensional space.

Schlegel Diagrams of Polyhedra

The central projections of regular polyhedra form beautiful patterns which reveal the structure and symmetries of these objects. If a regular polyhedron has all of its vertices on a sphere in three-space, then we may use central projection from the north pole to obtain an image on the horizontal plane at the south pole. For any vertex not lying at the north pole, the line from the north pole through the vertex meets the horizontal plane at the image of the vertex. If two vertices are connected by an edge in the polyhedron, then the image of the edge is the segment in the plane joining the images of the two vertices. The images of the edges of a polyhedron are said to form a *Schlegel diagram* of the polyhedron, named for Viktor Schlegel, the German mathematician who invented this type of diagram in 1883. Each polyhedron has many different kinds of images formed by central projection from the north pole, depending on how the polyhedron is rotated within the sphere. We will consider only Schlegel diagrams having a special property, namely that the image of the face closest to the north pole contains the images of all the other vertices.

The Schlegel diagram of the cube is the familiar "square-within-a-square." For a tetrahedron, the Schlegel diagram is a triangle with its vertices connected to its midpoint. To position an octahedron for its Schlegel diagram, we rotate the polyhedron so that two of the triangular faces are horizontal. In the resulting projection, the image of the top face is a large equilateral triangle, and the image of the bottom face is a smaller equilateral triangle rotated by a half turn. The images of the other six triangles of the octahedron are triangles in the plane, each joining an edge of the inner triangle to a vertex of the outer one or vice versa. In the corresponding diagram for an icosahedron, the 12 vertices are arranged in three nested polygons: a large equilateral triangle contains a regular hexagon, which in turn contains a small equilateral triangle. The Schlegel diagram of a dodecahedron has 20 vertices

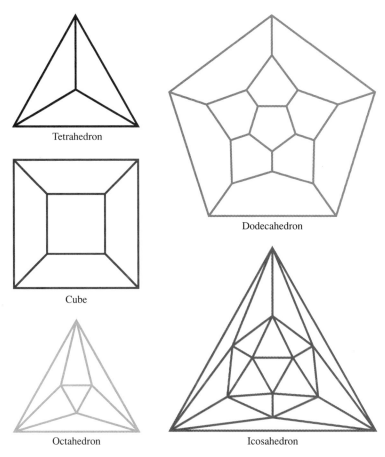

Tetrahedron

Cube

Octahedron

Dodecahedron

Icosahedron

Schlegel diagrams in the plane for the five regular polyhedra.

similarly arranged in three nested polygons: a large regular pentagon contains a nonconvex 10-sided polygon, which in turn contains a small regular pentagon. These Schlegel diagrams display in a nice way the symmetries of the regular polyhedra and their faces.

Schlegel Polyhedra for Regular Polytopes

We can create Schlegel polyhedra for the regular four-dimensional polytopes by means of central projection from four-space to three-space, the analogue of central projection from three-space to the plane. The Schlegel polyhedron of the hypercube is the cube within a cube, with corresponding vertices connected. Just as the Schlegel diagram of the tetrahedron is a triangle with its vertices

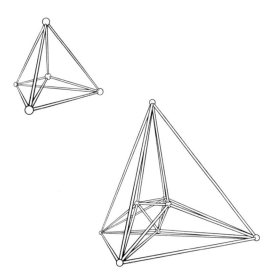

Schlegel polyhedra in three-space for the 5-cell and the 16-cell, illustrated in *Geometry and the Imagination* by David Hilbert and Stefan Cohn-Vossen.

connected to a central point, the Schlegel polyhedron of the four-simplex is a tetrahedron with its vertices connected to a central point. The six triangles joining this central point to the edges of the large tetrahedron divide the interior of the tetrahedron into four somewhat flattened triangular pyramids.

The Schlegel polyhedron for the 16-cell dual to the hypercube, composed of 16 tetrahedra, is similar in form to the Schlegel diagram of the octahedron dual to a cube. Instead of a triangle within a triangle, the Schlegel polyhedron of the 16-cell is a tetrahedron inside another rotated tetrahedron. Every vertex of the inner tetrahedron is connected to the three closest vertices on the outside tetrahedron, thus giving four more of the tetrahedra in the 16-cell, and four additional tetrahedra come from joining a vertex of the outer tetrahedron to the closest triangle of the inner tetrahedron. The remaining six faces are obtained by joining an edge of the inner tetrahedron to the closest edge on the outer one.

In the Schlegel diagram of the self-dual 24-cell, the vertices are arranged in three nested polyhedra: a large octahedron corresponding to the face closest to the viewing point, a small octahe-

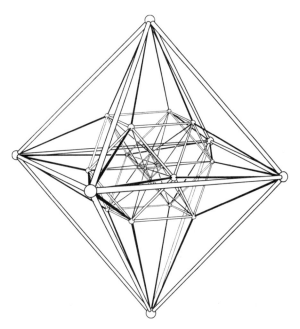

A Schlegel polyhedron in three-space for the 24-cell, also from Hilbert and Cohn-Vossen.

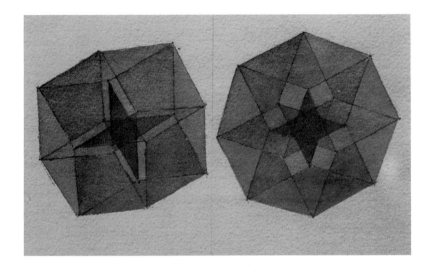

This watercolor by David Brisson shows two views of a hypercube in planes intersecting in a single point in four-space (see further explanation on pages 144 and 191), forming a "hyperstereogram," invented by Brisson. Only one portion of the figures will come together to form a true image in any view.

dron inside, and between the two a polyhedron called a *cuboctahedron*. The eight triangular faces and six square faces of the cuboctahedron are obtained by cutting off the corners from a cube all the way to the midpoint of each of the cube's 12 edges. Each of the eight triangles is a face of two octahedra, one with a triangular face on the outside octahedron and one with a triangular face on the inside octahedron. That accounts for 18 of the octahedra in the 24-cell. The remaining 6 octahedra are determined by joining a square face of the cuboctahedron to one vertex on the outer octahedron and one vertex on the inner octahedron. Similar presentations are possible for the 120-cell and the 600-cell, but their many vertices make the diagrams difficult to interpret. It is possible to construct stick models of the Schlegel polyhedra of the regular polytopes and to investigate them by turning them around in three-space. The models of Paul R. Donchian are the most famous physical constructions of such objects.

A single photograph of a Schlegel polyhedron can be quite confusing. As far back as the last century, mathematicians experimented with stereoscopic pairs of geometric objects, so that the left eye received one perspective image and the right eye received an image from a slightly different viewpoint, creating a three-dimensional effect. Although this technique is still useful in studying complicated configurations, the most effective way to see objects in three-space is to walk around them and record the separate images to form an animated film.

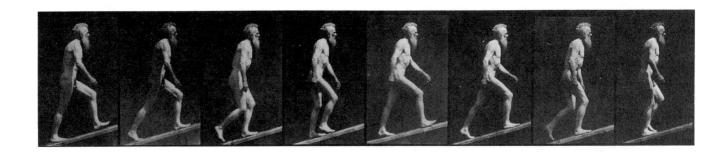

Eadweard Muybridge strides in front of his camera
to create one of the first animated films.

Animating the Hypercube

Within a generation of the invention of photography 150 years
ago, Eadweard Muybridge had used this new technology to alter
our perception of time and space. The photographs of Muybridge
himself, striding up a ramp in front of his camera, could be placed
on a rotary device and flipped over and over so that he walked on
and on, in a primitive version of a motion picture. Slow motion
and freeze frame techniques made it possible to analyze the motion of a racehorse or the exertion of muscles in lifting a log.

We can combine a century and a half of animation experience
together with modern computer graphics to create and investigate
complicated configurations in three-dimensional space. Architectural and industrial design become dynamic processes as we look
at not just a few views but 30 views per second, each slightly
different from its predecessor, giving the impression of continuous motion. We can experience what it would be like to walk
along a corridor or down a staircase in a building that has not yet
been constructed. As an architect takes her client on a tour of a
prospective auditorium, she can alter the different features to create different impressions. Should this window be placed slightly
higher? Should that entranceway be longer? A turn of a dial can
produce the new view and simultaneously make the changes for a
new set of blueprints.

Modern graphics computers can produce images very
quickly. For a wire-frame object, the computer calculates the positions of the vertices and draws the appropriate segments. The
speed of production and display depends strongly on the number
of vertices and edges in the object. Even on some relatively small
machines, a cube, with 8 vertices and 12 edges, can be rotated to

give the impression of continuous motion in what is called real time. The hypercube is not that much more complicated, with 16 vertices and 32 edges. A. K. Dewdney described ways of creating programs for hypercube rotation in the April 1986 "Computer Recreations" column of *Scientific American.*

To create an animation using parallel projection, the program keeps track of the position of one corner of the cube or hypercube and of all corners attached to it by edges. Once the images of these points are determined, all the other points and segments can be drawn easily since the images of parallel segments of the same length will be parallel segments of the same length (as described in Chapter 4). Some additional calculations are necessary for central projections since images of parallel segments will no longer be parallel but will lie in lines going through "vanishing points." The computer is quick enough to carry out these computations to produce frames in perspective of an animation as the hypercube rotates in four-dimensional space. Color-coding makes it possible to keep track of different parts of the rotating hypercube, as portrayed in the film *The Hypercube: Projections and Slicing.*

The second section of that film begins by showing central projections of the three-cube. In the first sequence, white edges join corresponding vertices of a red square and a green square lying opposite. As the cube rotates in three-space, its images under central projection change until at one point we have a green square inside a red one, then a red trapezoid next to a green trapezoid, and then a red square inside a green square. The analogous sequence for the hypercube (as it is shown in the illustration on the next page) starts with a red cube inside a blue cube joined by black edges stretching between corresponding vertices. The images formed when the hypercube rotates in four-dimensional space resemble those of the ordinary cube rotating in three-space. The large blue cube opens toward the top, flattens out, and opens inward to form an incomplete pyramid, while at the same time the small red cube opens toward the bottom to form another incomplete pyramid. If we continue the rotation, the blue cube will become the small cube, and the red cube will flatten out and come back to become the large cube.

In observing such a sequence not once but several times, the viewer gains an appreciation of the symmetry of the hypercube. Each of the eight cubical faces takes its turn holding all of the various positions in the Schlegel polyhedron. As each of the cubes flattens out and opens up again during the rotation, it changes

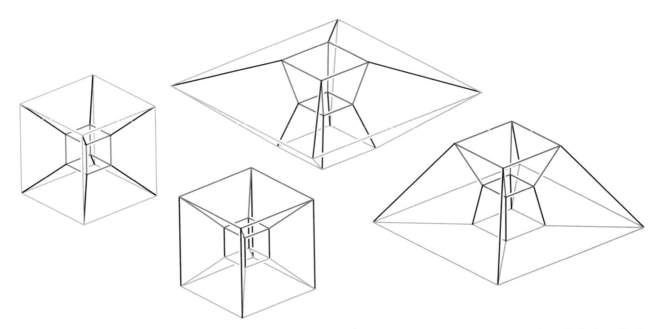

Clockwise starting from the upper left: a sequence of central projections of a rotating hypercube in four-space.

orientation. If a cube were to contain a right-handed glove before the flattening, the glove would have become left-handed afterward, and conversely. This reversal of orientation plays a central role in a famous philosophical dispute, which we will treat in Chapter 9. It also is a common source of ambiguity when we try to interpret moving visual images of objects from four-dimensional space. The more experience we have with understanding simple objects from several standpoints, the better is our chance of understanding images of new objects from four-dimensional space and higher.

The Polyhedral Torus in the Hypercube

Up to this time, we have been discussing images made of lines, the easiest objects for a computer to draw. As technology improved, computers were able to fill in polygonal regions, creating images better resembling solid objects. But which polygonal faces should be filled in?

When we draw a picture of a solid cube, we want only to see three of the six square faces, the ones in front hiding the back faces from view. As the cube rotates, different squares will come into view, and color-coding, or markings as on dice, can help to identify which particular view appears at any given time. For a cube or a regular polyhedron, it is not difficult to decide which faces should be filled in, although this task becomes significantly more subtle when rendering images of more complicated objects in three-space.

For a hypercube, deciding which squares to fill in can be a problem. We could fill them all in, but then when we project the hypercube into three-space, we will see only those images lying on the outside of a solid polyhedron. The edges and faces inside the object will be hidden. In the central projection, for example, when the red cube is largest, we would not see the blue cube at all. We could make different faces emerge from within by rotating the hypercube, or with a more sophisticated computer, we could arrange for some of the faces to be transparent.

An easier technique is to leave out certain of the 24 square faces. Instead of displaying 3 squares at each edge, we can choose 16 of the squares so that there are 2 of them at each edge of the hypercube, forming a two-dimensional surface in four-dimensional space called a polyhedral torus. We may think of this torus as having a fold-out pattern in the plane consisting of a square subdivided into 16 smaller squares, together with a set of instructions for assembling it. The instructions are to fold the object up so that the top edge can be attached to the bottom, which we can do in three-dimensional space to form a square tube. But the instructions also require that we fold the original square so that the left edge can be attached to the right. Once again we can accomplish this instruction in three-dimensional space. But in three-dimensional space, it is impossible to carry out both instructions without stretching or tearing the original subdivided square. It is, however, possible to accomplish this in four-dimensional space since we can choose 16 squares from the hypercube so that there are 4 squares at each vertex and 2 squares at each edge.

When we use central projection to produce an image of the hypercube in three-dimensional space and then show the image on the computer screen, some of the 16 squares have images that are squares, whereas others will appear to be trapezoids or more complicated four-sided polygons. All these polygons will fit together to form a boxlike approximation of the familiar shape of a torus of revolution, obtained by revolving a circle around an axis. As we rotate the hypercube in four-dimensional space, the changing images of the polyhedral torus in four-space will appear to

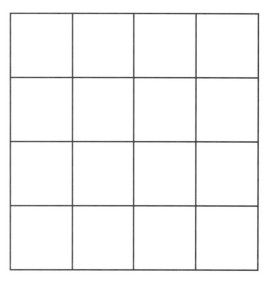

When attached top to bottom and side to side, a four-by-four grid forms the polyhedral torus in the hypercube.

The polyhedral torus in the hypercube.

swim through each other. To gain a better appreciation of the importance of this torus in the study of objects in four-space, it is useful first to consider central projection as used in cartography.

Stereographic Projection

Geographers have devised all sorts of projections from the curved surface of a sphere in three-space to flat two-dimensional maps. We have already been studying one of the most useful mapping techniques, central projection from a viewing point to a horizontal plane. When the viewing point is at the top of a sphere that rests on the horizontal plane, central projection sends each point of the sphere to a unique point of the plane. This gives a mapping from the sphere to the plane that cartographers call stereographic projection. To describe this mapping in terms of light rays, we begin with a transparent globe resting on a plane and imagine a bright light situated at the north pole. For each point on the sphere, some ray of light will pass through the point and create an image on the horizontal plane. Thus every point of the globe, other than the north pole itself, will have an image point on the plane, and every point of the plane corresponds to exactly one point on the sphere. The rays from the north pole through the points of the equator form a right circular cone, which cuts the plane in a circle. Similarly, the image of any parallel of latitude on the sphere is a circle in the plane.

This projection provides a very accurate map of the Antarctic region, but the map becomes more distorted near the equator. Land masses in the northern hemisphere are further distorted, and the image of Greenland for example is huge. To get a reasonable image of Greenland, we can rotate the globe, keeping the plane and the light source fixed, so that Greenland is moved near the point where the sphere touches the plane. Of course once this is done, Antarctica has moved very close to the source of the light rays, and its image is greatly distorted.

One of the significant properties of stereographic projection is that it not only sends parallels of latitude to circles—it sends *all* circles on the sphere to circles in the plane. Only circles passing through the north pole are exceptions; these have straight lines as their images. This property makes it much easier to keep track of what happens to the southern hemisphere as we rotate the globe around a horizontal axis through the center of the sphere.

Left: Central projection from the north pole sends a family of circles on a sphere to a family of circles on a horizontal plane. *Below:* Three views of central projection of the sphere to the plane as the sphere rotates through 90 degrees.

Before the rotation begins, the image of the southern hemisphere is the interior of a disc centered at the point where the sphere touches the plane. As we begin to rotate, the image of the equator is still a circle, moving off-center, and the image of the southern hemisphere is the interior of that off-center disc. Ultimately the equator rotates so far that it passes through the light source at the top point. The image of the equator is then a straight line, and the image of the southern hemisphere becomes an infinitely large half-plane. If we continue the rotation, the original

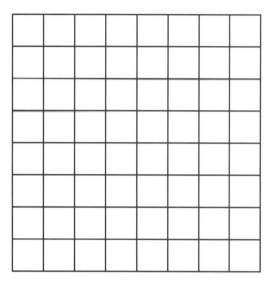

Eight-by-eight subdivision of the plan for the polyhedral torus.

southern hemisphere moves over the light source. Once the light source lies within the rotated southern hemisphere, the image of this region is the region outside the circular image of the equator. By the time the rotation has brought the equator back to a horizontal position, the southern and northern hemispheres have been interchanged and the images in the plane have been "turned inside out."

Stereographic Projection from Four-Space

We have described features of stereographic projection from the sphere in three-space to a plane. To describe this technique in the next higher dimension, we consider the effect of central projection on the analogue of a sphere in four-dimensional space, which we call a hypersphere. The ordinary sphere in three-space is the collection of points at a fixed distance from a given point. In four-space, a hypersphere is defined the same way, as the collection of all points at fixed distance from a given point. All vertices of a cube are at the same distance from its center, so the vertices of a cube are situated on a sphere about its center. Similarly, all vertices of a hypercube are situated on a hypersphere.

All of the vertices of the four-by-four polyhedral torus studied earlier in this chapter lie on a hypersphere since the vertices of this torus are the 16 vertices of the hypercube. In a similar way, we can assemble an eight-by-eight grid of 64 small squares in four-

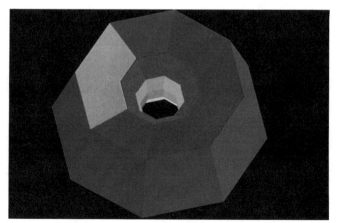

space so that all of its vertices lie on a hypersphere. We may use central projection to find the image of this polyhedral torus in three-space, and then look at its projections on the computer screen to see pictures of polyhedra in space which better approximate the smooth torus of revolution. Subdividing again and again produces a very close approximation of a surface in four-space having the smooth torus as its image under central projection. This particular torus, first studied in the last century by William Clifford, is known as the Clifford torus. This surface in four-space is of extreme importance in geometry and topology because of its remarkable symmetries, and it appears as well in the physics of dynamical systems, for example as a surface defined by equations governing the position and velocity of a pair of pendulums.

Central projection of the Clifford torus into three-space gives the familiar torus of revolution. As we saw in Chapter 3, the torus of revolution is covered by four different families of circles: the horizontal circles of latitude, the vertical circles of longitude, and two types of circles in planes making a 45-degree angle with the horizontal. Since central projection preserves circles on a hypersphere, the Clifford torus must also be covered by four families of circles. If we rotate the hypersphere containing the torus, keeping the light source fixed, then the images of the Clifford torus under stereographic projection become distorted, forming a collection of objects called cyclides of Dupin after their discoverer, Claude Dupin.

We can display the structure of the Clifford torus in four-space by dividing the surface into cylindrical bands and removing

Central projections into three-space of successive subdivisions of the polyhedral torus. As the number of subdivisions increases, the polyhedral torus more closely approximates the Clifford torus in the hypersphere.

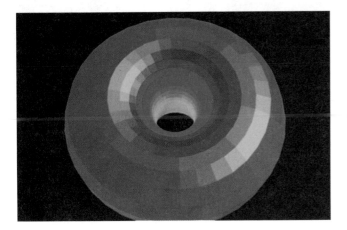

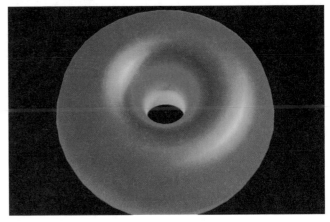

half of them. As we rotate the hypersphere in four-space, some portions of the torus have more accurate images and other parts become distorted. If we rotate so that one of the points of the torus goes through the light source, then the image in three-dimensional space stretches out to infinity. This infinitely large surface is remarkable in that, like the plane, it totally and symmetrically separates all of space into two congruent pieces. As we continue the rotation in four-dimensional space, the image returns to a familiar

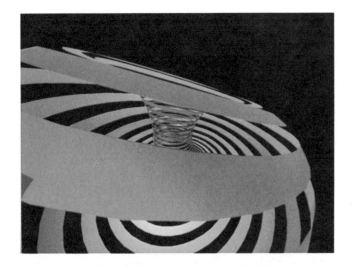

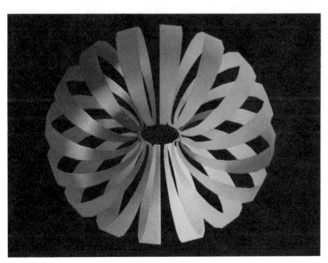

Seven views of central projections of the Clifford torus as the figure rotates in four-space. The bands that appear to run horizontally around the torus at the start appear to be vertical after one-quarter rotation.

torus of revolution but with an essential difference: the cylindrical bands that formerly were parallels of latitude have now become meridians of longitude. During this deformation, the torus has turned completely inside out! Latitude and longitude have interchanged.

In the next chapter we will use this central projection of a torus on the hypersphere to model the motion of a pair of pendulums.

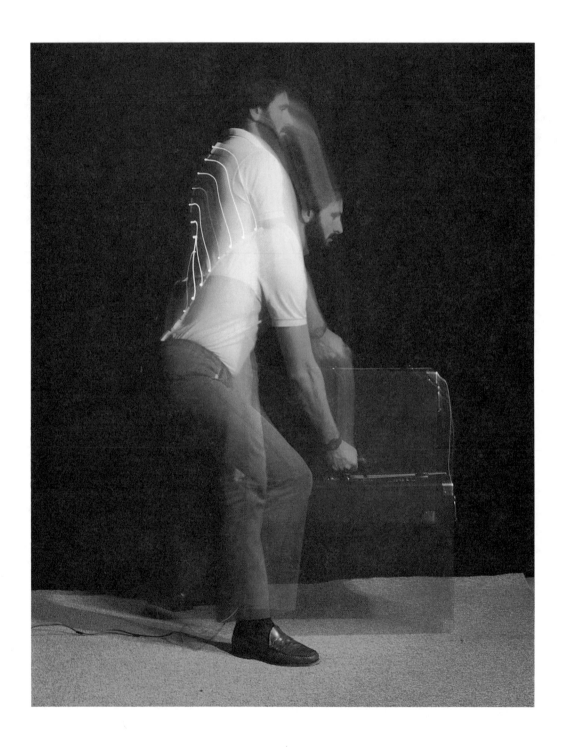

7 CONFIGURATION SPACES

At the Vermont Rehabilitation Engineering Center of the University of Vermont, Gerald Weisman and his associates research one of the most widespread and costly medical complaints, lower back pain. Workers in all types of jobs must carry out routine tasks that exercise their back muscles in different ways. An injury can make those tasks impossible to perform, and it is the goal of the rehabilitation engineer to determine when the injured workers have recovered sufficiently to return to work. It isn't enough simply to measure general strength or endurance. The doctors need to obtain an accurate and useful description of the physical demands of a particular job in order to tell whether or not a worker is in condition to meet them. What positions do workers assume in the course of their jobs? How long do they stay in these positions, bearing what kinds of load? How frequently do they carry out different sorts of bending and twisting? This analysis often becomes an exercise in dimensionality, and the visualization of data in different dimensions is the common thread in the examples of this chapter.

A goniometer strapped to the back of a worker records the different angles of bending and twisting during a lifting task. The collection of such positions is an example of a configuration space.

The Dimensionality of Rehabilitation Therapy

To study the range of possible positions, researchers attach to a worker a device called a goniometer, which records back position by keeping track of three angles. Two angles give the position of the spine, in coordinates roughly like latitude and longitude. The third coordinate measures the angle of the sideways twist of the shoulders relative to the pelvis. These three numbers describe a configuration that is the position of the device (and its wearer) at any instant during the course of the job. If we display these coordinates on a three-dimensional grid, then any single configuration corresponds to a point on this grid. We call this collection of points in three-space the configuration space for this particular example. As a worker moves from one position to another, the corresponding point moves from one location to another in the configuration space. As a worker moves through a whole series of tasks, we obtain a series of points tracing out a path in the configuration space describing the physical demands of the job.

The record of such a path in a three-dimensional space does not describe the actions of the worker completely. There is no information here that tells whether a person has walked to another location, or climbed a ladder, while bending and twisting. Only the bending and twisting are recorded since those are the significant variables in assessing the condition of the lower back.

Each job has its own dimensionality, depending on the number of different directions of bending and twisting the job involves; the more directions, the higher the dimension of the configuration space where we keep track of the possible positions a worker assumes. A park attendant collecting pieces of paper bends over many times, always in the same direction, and the goniometer registers the changes in this one and only angle. It would be possible to describe the bendings that take place over an entire day on a sort of seismograph, producing a strip of paper showing the angle reading at each time. One could then easily determine how often the park attendant bent over past a certain angle, and for how long. In terms of the goniometer, the park attendant has a one-dimensional job.

A secretary may bend from side to side as well as bend forward. His positions can then be specified by two coordinates. Since there is no twisting, the third coordinate angle is always zero, and the state of his lower back at any particular time can be

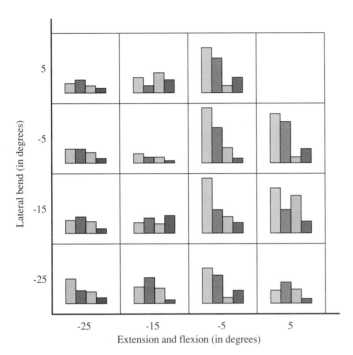

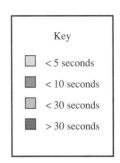

In this graphing of goniometer readings the height of the bars indicates the amount of time spent in bending and flexing during the performance of a task.

indicated by a point on a two-dimensional grid. As he goes through the motions of his job, the recording point could trace out a path on a two-dimensional computer screen, an "orbit" in the configuration space. Note that in this example there is no time axis. If we wanted to know at what time the secretary assumed a given position, we would have to label the corresponding point on the two-dimensional record sheet.

Researchers at the University of Vermont analyze such orbits by dividing the grid up into cells. By noting the number of times an orbit curve enters a particular cell during a workday, they can measure the complexity and demands of the job. They can then determine which kinds of orbits are suitable for workers with different injuries.

For a job that requires twisting as well as two different kinds of bending, each goniometer reading has three coordinates, and the orbit curve will lie in a three-dimensional space. Once again, the technique is to divide the configuration space into cells or "bins" and to see how many times the orbit curve enters different bins in the space. As we have seen in previous chapters, analyzing a geometric object in three-space often involves projecting it into two-dimensional planes, and that is what occurs here as well.

Not all the variables in lower back research can be described in a three-dimensional configuration space. The rehabilitation therapist might wish to consider additional variables affecting the lower back, like the weight of load to be lifted or the temperature of the environment, so the appropriate configuration space might have more than three dimensions. To analyze a collection of points in a configuration space of four or five dimensions, researchers make use of the geometric techniques similar to those that we have developed for analyzing fundamental objects like hypercubes. Such higher-dimensional structures can form the framework for the understanding and interpretation of data from many different fields.

Dimensionality and Dance

Professor Julie Strandberg of the Program in Dance at Brown University explored freedom of movement by choreographing *Dimensions*, a twenty-minute piece for two dozen dancers that has had several performances in two different years at Brown University and once in New York City. As preparation for participating in this work, dancers began with a series of exercises designed to increase their appreciation of the effect on their movements of limiting their dimensionality. They lay down on their backs or stomachs or sides on the gymnasium floor and tried to wriggle along while staying in the second dimension. Standing up, they tried to walk along a line "Egyptian hieroglyphic style," always staying as much as possible in a vertical plane. They explored the possible motions they could make with their backs up against a wall—for example, they might be able to do cartwheels but not somersaults. For one dancer to pass to the other side of a fellow dancer while remaining flat against the wall is an athletic challenge.

In the actual performance, the trained dancers eventually leave the confines of the wall, first to move in increasingly complicated ways while maintaining contact with the two-dimensional stage, and ultimately moving beyond the limitations of

Dancers in *Dimensions* simulate the movement of polygons constrained within a vertical surface.

gravity with leaps and lifts and movement on swings. The dance tells the story of a Flatlander who is suddenly introduced to the world of three dimensions. The contrast between the extremely constricted movement of even the most imaginative Flatlanders and the freedom of motion of the Spaceland dancers is striking. The sole Flatlander to be given a glimpse of this higher-dimensional paradise is entranced by the movement of a pair of dancers, whom she joins by inserting herself, like a playing card, between them until they help her to move in space by herself. There is a real sense of liberation, heightened by the use of color and intricate rhythms to accompany the twists and turns and leaps and lifts that are not possible in the planar environment. As it happens, the experience is too much for her, and all too soon she plummets back to her Flatland plane, racked by the memory of that freer, higher-dimensional land. It is a sad tale, but a magnificent parable.

The configuration space of all possible dance positions has extremely high dimensionality. The position of the left upper arm relative to the shoulders is already three-dimensional, determined by the same sort of angles that are measured with a goniometer in the analysis of the lower back. The position of the lower arm, rotated and twisted relative to the upper, adds another three dimensions. We have six dimensions for the left arm and we have not even reached the wrist! Only when we record and analyze such configurations do we begin to realize the dimensionality of the world we move through without thinking.

A climber scales a sheer cliff in a virtually two-dimensional environment.

Orbits of Dynamical Systems

A much simpler configuration space, of only four dimensions, arose in the work of two researchers in the Division of Applied Mathematics at Brown. Professors Hüseyin Koçak and Fred Bisshopp generated enormous lists of numbers from their experiments with a pair of pendulums. A computer simulated vast numbers of positions and velocities for the pair of pendulums, depending on different starting points and different ratios between the frequencies of the pendulums. Koçak and Bisshopp wanted to display their data in a visual way that would reveal patterns impossible to see in a mere catalogue of numbers. The applied mathematicians came to our geometry project to see if our methods for visualizing higher-dimensional configurations could help them in their work. As it happened, the fit between our two research projects could not have been better.

Koçak and Bisshopp recorded the position of each pendulum by marking a point on a circle in a two-dimensional plane, so at any given time points on two circles specified the positions of the two pendulums. Since each observation involved two points, each with two coordinates in the plane, the data set was four-dimensional. The two circles could have been graphed side-by-side on a plane, but that representation would not have indicated sufficiently how the motions of the two pendulums were related. A new technique was needed for displaying the relationships more effectively.

We can see the relationship between two variables like height and armspan by graphing them on a plane, thought of as a line of lines, with one vertical line for each point on the horizontal axis. Similarly, we can see the relationship between points on circles by graphing them on a torus, thought of as a circle of circles. We could draw the graph of such a relationship on a torus in three-dimensional space, with one vertical circle for each horizontal "axis circle." But we can get an even more symmetrical graph if we use a four-dimensional configuration space, plotting the positions of the pendulum system on the Clifford torus on the hypersphere described in the previous chapter. This more symmetrical representation made it possible to recognize the most significant patterns in the different orbit curves.

The history of the motion of the two pendulums could be represented as a sequence of points tracing out a curve on the

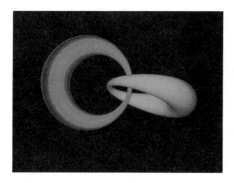 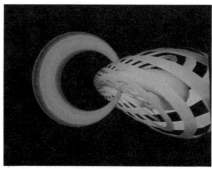 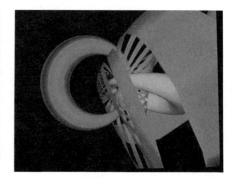

torus in the hypersphere. Visualizing the structure of the orbits for different choices of beginning positions and frequency ratios of the two pendulums was a challenge that was well suited to our computer graphics techniques. The way the data points from the physical experiment were arranged in nearby orbits suggested displaying the surface in strips, a device that turned out to be especially effective for the presentation and investigation of other surfaces in three- and four-dimensional space.

When one of the pendulums is stationary, the orbits traced on the torus are circles of latitude or longitude. When both pendulums move in a synchronized way, beating in unison, the orbit curve goes around the torus once each way, hitting each parallel of latitude and each meridian of longitude exactly once. These orbits turn out to be precisely the circles that we obtained in Chapter 3 by slicing a torus obliquely so that the slicing plane is tangent to the torus at two points. The film *The Hypersphere: Foliation and Projections*, made together with Koçak, Bisshopp, and computer science graduate students David Laidlaw and David Margolis, presents a visualization of all possible orbits of synchronized pendulums, named *Hopf circles* after the Swiss mathematician Heinz Hopf, who studied their properties in the 1930s. The collection of Hopf circles on the hypersphere is one of the most intriguing higher-dimensional images. More complicated data sets produce orbits of greater complexity, leading to knotted curves and curves that do not close up after a certain amount of time. It is by comparing such complex systems to the collection of Hopf circles that researchers can begin to visualize more subtle relationships in the orbits of dynamical systems.

A sequence of images showing the orbits of synchronized pendulums as the edges of blue/gray bands on torus surfaces. The frequency of a pendulum is determined by its length; each torus contains orbits for pendulums of two particular frequencies, hence two particular lengths. Each surface contains orbits for pendulums whose lengths have a fixed ratio, so that the purple torus is filled with orbits where the first pendulum is much shorter than the second, and the blue-green torus corresponds to those where the second pendulum is much shorter than the first. The sequence of blue/gray bands shows these surfaces as the initially shorter pendulum becomes longer and the other pendulum becomes shorter. Halfway between blue-green and purple rings is a surface stretching to infinity containing the orbits of pairs of pendulums with the same length.

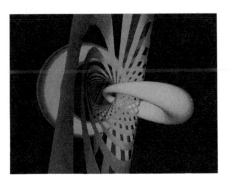

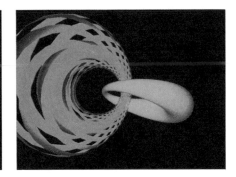

Anthropological Sites and the Space of Circles

Lower-dimensional configurations of circles arise in fields far from physics and mathematics, and a geometric understanding of dimensions can help there as well. Professor Richard Gould in the Anthropology Department at Brown University found a specific use for the configuration space of circles in the plane while organizing data about hunter-gatherers in the Australian bush. A colony of hunter-gatherers can maintain a fixed location as long as its hunters can find a supply of game in all directions. From the central campfire, hunting parties spread over a large area, which can be considered as a disc with a given radius. If large predators threaten, hunters will return each night to a secure camp and the size of the disc will be relatively small. In the absence of major threats, they can stay away for one or more nights before returning to camp, and the area of the disc centered at the campsite will be larger. Ultimately the campsite will be moved. Gould wanted to know how far a colony would move, and what the pattern of movement is over time.

To keep track of a number of circles, Gould recorded each one with three numbers, using latitude and longitude to give the posi-

Aboriginal hunter-gatherers at a desert camp near Tikatika, Western Australia.

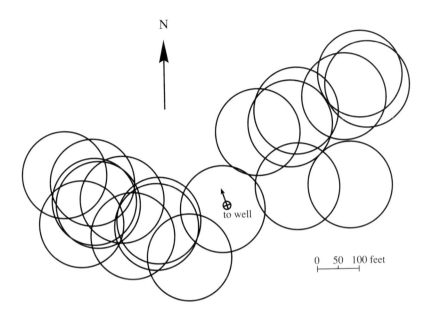

N

to well

0   50   100 feet

Overlapping circle patterns show the location of campsites and the areas covered by hunting parties near the Mulyangiril well in Western Australia.

tion of the campfire and a third number to give the radius of the disc. The space of discs is then three-dimensional, and the movement of the colony determines a sequence of points in this space, a "polygon in the space of circles." To find the total area of the campsites over time requires knowledge of the ways the discs overlap, and here the three-dimensional recording system has an additional advantage—for each pair of sites, Gould could compute a number telling whether or not they overlap, and in particular whether or not one site is completely contained in another.

As it happens, the mathematics of this problem has a curious history. The space of circles was studied as a mathematical subject in the last century by the French mathematician Edmond Laguerre. He thought of the triple representing a given disc as a point in ordinary three-dimensional space, and the condition for two discs to be intersecting, or for one to contain the other, was given by attaching a "distance" to a pair of points in space. But instead of using the ordinary Pythagorean theorem to represent the distance in terms of a sum of squares, Laguerre's distance involved the difference of squares. It was precisely this sort of generalized distance that turned out to be significant in the geometry of relativity. We can develop this notion further by investigating a less realistic configuration space, the space of spotlights on a stage.

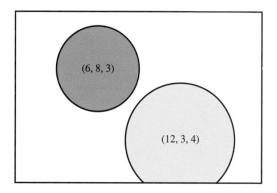

Cones of light from bulbs suspended above a stage determine circle patterns on the floor.

The coordinates for the center and radius for each disc of light indicate whether or not the disc will remain completely within the stage area.

The Dimensionality of Stage Lighting

Consider a highly simplified example that leads to a fairly intricate space of geometric objects. The lighting director of a local theater has to arrange a set of lights over the stage so that they illuminate certain parts of the floor at certain times. Sometimes the size of a spot needs to change during the course of a performance. Sometimes one colored circle of light must be contained in another. How can she keep track of all the circles of light, and how can she design her lighting directions so that her assistants can carry them out?

The lights all have the same form. A single bulb is suspended from a wire hanging from the ceiling, and a conical shade directs the light out in a beam, which meets the floor in a disc of light. The sides of the shade come down at such an angle that the radius of the disc is equal to the height of the bulb above the floor. Thus the lighting director can specify the radius of the disc by specifying the height of the bulb above the floor, a feature that makes it easy for the director to specify the location of any light. She can always indicate the position of the center of the disc by noting the coordinates used by the director of the play to give her instructions, and for a complete specification, she adds a third coordinate giving the radius, or the height. The collection of lights is thus a three-dimensional configuration space.

On one level, the use of three coordinates provides a convenient way to record the various lights. For example, the coordinates (6, 8, 3) might indicate a light centered at a stage point 6 feet from the left side and 8 feet from the edge of the stage, having radius 3. On a deeper level, the collections of coordinates define a geometrical space. When mathematicians call a collection a space, that term generally indicates that more structure is present. In the case at hand, it is possible to use the coordinates to determine particular properties of the lights and their relationship to the stage and to one another. For example, the light with coordinates (6, 8, 3) stays on the stage, while the light corresponding to (12, 3, 4) comes off the front of the stage. It is easy to determine a rule that tells when a light stays away from the front rim of the stage, namely that the second coordinate is larger than the third. In this way we see a relationship between the geometry of the configurations and the relationships among their coordinate representations.

The lighting director can also solve more complex problems by referring to the coordinates. For example, when will one spot be contained in another? This happens when the distance between the points in the plane given by the first two coordinates is greater than the difference of the third coordinates. In this space, the three coordinates do not play the same roles, so even though the geometry of the configuration space is three-dimensional, it is not identical with the usual geometry of ordinary three-space.

This example sheds light on the introduction of time as a fourth coordinate. Sooner or later everyone hears a statement like "Time is the fourth dimension," and that idea represents a limitation of the idea of dimensionality. Already in the last century writers realized that in many situations time can be viewed as *a* fourth dimension, but by no means does it demand any special role as *the* fourth dimension. When physicists, especially relativity physicists, specify an event by giving three space coordinates and one time coordinate, they are using a four-dimensional configuration space. This space has its own geometry, but it is not the same as the geometry of the four-dimensional space that extends ordinary plane and solid geometry, where distance is given by the generalized Pythagorean theorem. In relativity, the distance between two events is given by $\sqrt{(x - x')^2 + (y - y')^2 + (z - z')^2 - (t - t')^2}$, so the time coordinate, measured in special units related to the speed of light, appears in the expression with a negative sign, not a positive sign as in the generalized Pythagorean theorem.

The three-dimensional configuration space of spotlights is analogous to a four-dimensional geometry occurring in molecular modeling. The atoms making up a molecule are modeled by small spheres of varying radii. The description of the arrangement of atoms in a particular molecule consists of a list of such spheres, each with three coordinates to specify its center and one coordinate to give the radius. Thus the configuration space of atoms is four-dimensional. We can describe such a molecule to a graphics computer and ask it to display any particular view of the object. The computer can determine that two atoms do not intersect by checking an algebraic condition in four coordinates, namely $(x - x')^2 + (y - y')^2 + (z - z')^2 - (r + r')^2 < 0$. The geometry of this configuration space is much closer to that of relativity theory than it is to ordinary Euclidean four-dimensional geometry.

The lighting system that began our discussion can be even more complicated if each of the lights also possesses a rheostat to

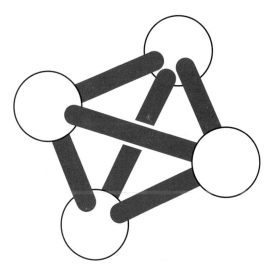

A simple molecule is modeled by a collection of nonoverlapping spheres joined by rods.

control the current, and hence the brightness of the spot. Adding the coordinate for brightness causes the configuration space to become four-dimensional. If we want to indicate the color of each spotlight as well, then the dimensionality jumps again. Usually we describe a color using separate numbers for hue, saturation, and value, or using three numbers giving the relative amounts of red, yellow, and blue (for pigments) or red, green, and blue (for lights). In any case the specification of color requires three more coordinates, so the lighting director will have seven coordinates for each spotlight—two for floor position, one for radius, one for brightness, and three for color. Thus it is that a simple example leads to a configuration space of high dimensionality.

The universe of modern physics is much more complicated than Einstein's description of events based on three dimensions of space and one of time. Some current models keep track of ten dimensions that act like space and one that acts like time to give an 11-dimensional configuration space. Another important model uses a configuration space with 26 dimensions. In each case the choice of the model depends to some degree on the kinds of mathematics that apply in these dimensions, as an aid for keeping track of the complex interrelationships among events in these higher-dimensional spaces.

Configuration Spaces of Segments and Lines

The geometric space of segments is a configuration space with a long history. In this geometry, the fundamental elements are not points but segments, determined by pairs of endpoints. Already studied in the last century as an example of a real four-dimensional geometry, the geometry of segments can describe architectural structures made up of straight boards as well as works of art created by sculptors such as Naum Gabo by stretching strings between frameworks. The fact that the separate pieces of many of Gabo's works can be described by giving simple formulas adds a new element to their beauty, and as we understand how to develop basic shapes in the space of segments, we can appreciate even more the process of artistic creation that assembles these components in such powerful form.

The complexity of the final design often reflects its dimensionality. Consider the following stylized account of a progres-

Naum Gabo's sculpture *Linear Construction in Space No. 1.* Nylon threads are stretched on a lucite frame, creating a structure in the configuration space of segments.

sively more complicated set of challenges leading to an important four-dimensional geometry of segments in space. For a sculpture show, two artists decide to decorate a wall with a pattern of plastic strings. They come up with a pleasing design by stretching 20 strings from the left-hand edge of the wall down to the baseboard. So that they can put the strings up again later, they have to find a way of recording the string positions. By recording two numbers, they can specify the position of any string. For example, the pair of numbers (4, 3) indicates the string that goes from the point 4 feet over along the floor to the point 3 feet up on the wall edge. Since it takes just two spots to place a given string, the dimensionality of the configuration space is evidently two.

In a way, constructing a sculpture in the space of segments is like the old game of "connect the dots." In the plane, a polygon is determined by a sequence of ordered pairs, and by connecting the dots in order, we draw the polygon. In the example at upper right, the basic elements are not points but segments, and we are recording a "polygon of segments."

We can increase the dimensionality of the collection of strings by allowing the bottom of a string to be placed anywhere on the floor, leaving the top still somewhere on the left edge of the wall. We still need one number for the height, but now the record will include two more numbers for the floor coordinates. The collection of segments is now three-dimensional.

By allowing the strings to start anywhere on the vertical wall and end up anywhere on the floor, we have an actual four-dimensional system. By convention we could specify each line by four coordinates, the first two giving the floor endpoint, and the third and fourth giving the baseboard and height coordinates for the wall endpoint.

An example of a "curve" in this geometry might be obtained by connecting points on a vertical line on the wall to a line parallel to the baseboard, moving along the baseboard by a fixed amount each time we move down the upright by a fixed amount. This gives a sequence of segments in space corresponding to a sequence of points in the configuration space. In the configuration space, these points lie along a straight line. In space, the sequence of segments all lie on a familiar architectural element, a hyperbolic paraboloid.

From the space of segments we move to the space of lines. Any segment determines a line, and in fact for each line meeting a pair of planes in distinct points, we find a segment determined by

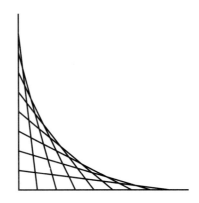

Two points moving uniformly along coordinate axes determine a two-dimensional geometry of segments.

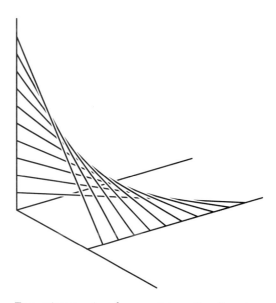

Two points moving along nonintersecting lines in space determine a hyperbolic paraboloid, formed here as a collection of strings.

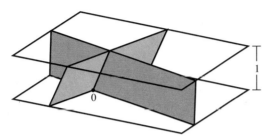

Each line in the horizontal plane at height one corresponds to a plane through the origin in three-space.

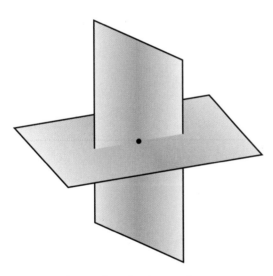

Four coordinates describe the set of two-dimensional planes through the origin in four-space. Two parallelograms centered at the origin in four-dimensional space may intersect in a single point. In this three-dimensional diagram, points are colored to indicate their height in four-space, so there is just one pair of points with the same fourth coordinate as well as the same first three coordinates.

the points of intersection. Thus we may specify the lines meeting the two planes by four coordinates, two for each intersection point, and the space of lines in three-space is four-dimensional.

Calculations can tell us whether or not two strings intersect each other. It is commonplace for strings laid out along a wall to intersect. Such intersections are rare if we choose the lines from among a three-dimensional collection, and rarer still for the four-dimensional system of lines in space.

Whereas the collection of lines in three-space is four-dimensional, the collection of lines in the plane is two-dimensional since we can specify a line not through the origin by telling the points where it intersects the two coordinate axes. This collection of lines in the plane is also related to the collection of planes through the origin in space. To see this, start with a fixed plane not through the origin. Then for almost any plane through the origin, the intersection with this fixed plane is a line. This sets up a correspondence between planes through the origin and lines in a plane, so the set of planes through the origin in three-space is two-dimensional. Similarly, the set of two-dimensional planes through the origin in four-dimensional space is in one-to-one correspondence with the lines in a three-space not going through the origin. This correspondence is at the heart of the subject of projective geometry.

In the spaces of lines or planes, there always seem to be special cases: lines or planes not specified by any choice of coordinates. Consider for example the two-dimensional space of lines in the plane. If we identify a line by its intersections with a horizontal and a vertical axes in the plane (the "intercept form"), then we miss those lines that pass through the origin. If we identify a line by its intersections with two vertical lines (the "slope-intercept form"), then we miss all other vertical lines. Similarly, if we identify lines in space by their intersections with two planes, then if the planes intersect, we miss the lines passing through their intersection line, and if the planes are parallel, we miss all lines lying in other parallel planes. If we are interested in the geometry near any given line, we can choose the reference planes to avoid any problem identifying nearby lines, and in this way we can study the entire projective geometry of the space of lines or planes.

We encounter a similar difficulty finding a valid coordinate system for the entire sphere. In the standard latitude-longitude system, we cannot assign unique coordinates to the north and south poles, where longitude lines converge. These are called the singular points of the chart. If we rotate the sphere while leaving the coordinate lines fixed, then there will be no singular points in the Arctic and Antarctic regions, but some other points of the sphere will now be singular. We do know that we can obtain an *atlas*, a collection of charts such that every point is a nonsingular point for at least one chart, and such that there is enough overlap between charts so that we can plan a route from any point to any other point. This notion of an atlas of charts is at the heart of the definition of a particular sort of configuration space called a *manifold*. In this type of space, every point lies in a region having a chart without singularities, and there is sufficient overlap to allow us to compare the geometry at one point to that of another.

Once we set up an atlas of charts on a surface, we can tell when a function on the surface is *differentiable*, a term indicating that the function can be approximated by the graph of a linear function, producing a line or plane or hyperplane, depending on the dimension. The collection of all differentiable functions is an extremely important feature of the surface, called its *differentiable structure*. It was found rather early that there was essentially only one possible differentiable structure on the ordinary two-dimensional sphere in three-space, and mathematicians suspected that in general there would only be one differentiable structure on a sphere of any dimension. It was quite a surprise therefore when John Milnor found in 1958 that there were essentially different

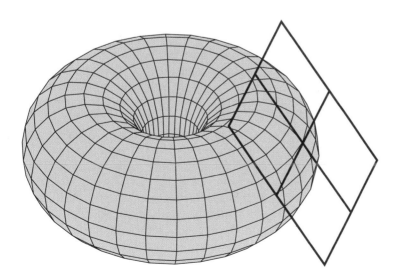

A graph of a function describing the surface of a torus in three-space is approximated by the plane tangent to the surface at one of its points.

ways of putting a differentiable structure on the seven-dimensional sphere. It was possible to construct an atlas with a perfectly consistent set of charts, so that one could identify which functions were differentiable, and to compare this collection with the differentiable functions on the ordinary seven-sphere in eight-dimensional space. The two collections were different. Milnor had constructed what he called an *exotic* differentiable structure on a higher-dimensional sphere, thereby opening up an entire field of study, *differential topology*.

In spite of the existence of exotic structures on the sphere, mathematicians generally believed that at least the differentiable structure of ordinary space was uniquely determined. This fact had been proven for all dimensions except four, and mathematicians expected that this too would be proved in due course. It was astounding therefore when in the early 1980s the work of two young mathematicians, Michael Freedman and Simon Donaldson, combined to show that the common belief was false—there are infinitely many ways to construct a differentiable structure on four-dimensional space.

Wave Fronts and Focal Curves in the Plane

Heaters and loudspeakers affect their surrounding space by radiating waves of heat or waves of sound. The nature of the waves depends on the shape of the object causing them and on its position in space. Many wave sources have the property of focusing waves the way a lens focuses waves of light. Two overlapping branches of mathematics known as geometric acoustics and geometric optics both study the propagation of wave fronts and the focusing properties of curves and surfaces, and in doing so, they bring together the geometry of lines in space and the geometry of circles in the plane.

Waves radiating from a source often interfere with each other, creating focal points. The collections of focal points from the various waves form intricate patterns, increasing in complexity as the dimension of the wave source increases. Once again the dimensional analogy helps us to appreciate these patterns. With a firm understanding of the geometry of focal points for curves in the plane, we are much better able to grasp the far more complicated surfaces of focal points in three-dimensional space. And often it is possible to understand the geometry of a wave source in one space by looking at a higher-dimensional configuration space of lines or circles associated with the object.

The wave patterns are simplest when the source radiating them is the simplest possible object, a point. When we speak into a telephone, our voice acts as a point source causing waves to travel out along the wire. Similarly when we heat the tip of a metal rod, heat travels out along the rod. In each case, a single "point wave" is moving along a one-dimensional space. By coloring each point according to its temperature, we obtain a spectrum along the path of the wave indicating how far it has traveled, from red-hot down to a cold violet. Heating the midpoint of a metal rod sends a heat wave traveling out in both directions. Points equally distant from the source of heat will have the same temperature and therefore the same color.

Moving up to two dimensions, we consider ripples from a pebble tossed into a still pond, or a brushfire spreading out from a lightning strike. The wave fronts will be concentric circles, expanding out in a plane from a single point source. If the point source radiates heat, then points of equal temperature are collected into "isothermal circles," like the isotherms on a weather map, which also connect points of equal temperature. By coloring each circle according to its distance from the origin, we in a sense transform one variable into another.

If the source of heat is a baseboard radiator along a wall, then the points on the floor with the same temperature will lie in lines parallel to the wall. Heat radiates outward in straight waves that never interfere with one another. The lack of interference is a manifestation of the straightness of the line source of the wave.

More interesting patterns emerge if the source is not a point or a line but rather an object with curvature. The simplest example occurs when a wave travels out from a circle in the plane. If we strike a metal circular cylinder immersed in a still pond, we send a pair of shock waves moving along the surface of the water. One shock wave merely continues outward in an ever-expanding set of nonintersecting concentric circles. The other wave moves inward toward the center. At the very center, the wave shrinks down to a single focal point and reemerges as another circle with the same center, heading outward and eventually passing through the cylinder. The waves will cover each point other than the center exactly twice. The color coding is not quite so effective as in the case of a point source or line source since it is no longer true that each point has a unique color.

Concentric circles radiate out from a point source.

Circular waves emanating from points of a circle converge at the circle's center and reemerge rotated 180 degrees.

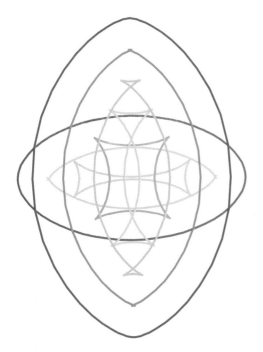

An ellipse and several of its interior parallel curves.

We can use color in another way to indicate what is happening as the inner circle shrinks down to a point and reappears. If we color each point of the original circle according to the usual color wheel, then the points are labeled red, orange, yellow, green, blue, and violet as we move counterclockwise around the circle. As the wave front moves inward from this circle, each point moves in a straight line. After these colored lines come together at the focal point and then reemerge, the red point is directly opposite its starting point, to the far left rather than the far right. Each colored point on the wave front is directly opposite its starting point, so the circle has turned itself inside out in the process of passing through its focal point.

The behavior of waves is far more complex when they originate from a curve having varying curvature, like an ellipse. To represent the region swept out by the parallel waves shortly after they leave the ellipse, we draw circles of the same small radius centered at the points of the ellipse. The boundary of the region covered by this collection of discs is a pair of parallel curves to the ellipse. One parallel curve continues outward, yielding a family of curves that never interfere with one another. They appear nearly elliptical in shape, although they are not exactly ellipses.

The inner parallel curve on the other hand behaves very differently. Shortly after it moves in toward the center, the parallel curve develops *singularities*, cusplike points where the curve abruptly changes its direction. Although in the case of a circle, an entire parallel curve collapsed to a point in one instant, in the case of the ellipse, the focusing takes place at different times. The first singularities appear near the most tightly curved points of the dark blue ellipse. At one instant, the innermost light blue curve appears to have two sharp-angled corners. These corners immediately spawn a pair of "fishtails," so that the parallel curve now has four sharp *cusp* points and a pair of double points where the parallel curve cuts through itself. As the parallel curve continues farther, the double points coalesce to a single point (light green) and then disappear, leaving a dark green curve with four cusps and no crossing points. A bit farther along, two arcs come together to form a parallel curve having a pair of double points and four cusps (yellow and light orange). The two pairs of cusps then merge with the double points (dark orange), and after this stage, the parallel curve once again becomes a singularity-free red curve resembling an ellipse. During this process, the ellipse has completely turned itself inside out.

If we display a collection of the parallel curves all on the same diagram, we see a new phenomenon that might be difficult to identify if we only saw the individual parallel curves: the cusps of the curves parallel to the ellipse trace out another curve, the *focal curve* or *evolute* of the ellipse. If we know a curve well enough to generate its parallel curves, then we may obtain the focal curve as the collection of all cusp points on all parallel curves.

We get a new insight into the geometry of this important focal curve by changing our viewpoint. Instead of thinking of parallel curves emanating all at once from an original curve, we may think of rays perpendicular to the curve proceeding from a number of points spaced out along the curve. We can imagine each of the patrons seated at the rim of an ellipsoidal stadium pointing a laser in the direction straight out from the rim. Viewed from a blimp high above the stadium, the light rays would interfere with each other to create a bright curve called the *light caustic* by researchers in mathematical optics. It is identical with the focal curve.

In the case of a circular stadium, all of the rays coming out from points of the rim will coalesce at a single point, and the evolute degenerates to a single point. If the circle deforms to an ellipse, this degenerate evolute opens out into a curve.

The characteristics of the evolute curve tell us something about the way that the curvature changes as we go around the ellipse. For example, the evolute has the same symmetries as the original ellipse. It has its own singularities, two cusps on the

Left: When a large number of parallel curves of an ellipse are displayed together, their cusps trace out the focal curve of the ellipse. *Right:* Rays coming straight out from the points of an ellipse show the focal curve as a light caustic.

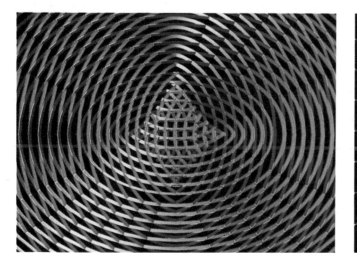

major axis of the ellipse and two cusps on the minor axis. The rays coming from the points of the ellipse cover some parts of the plane twice and other parts four times. The evolute curve separates the region of twofold coverage from the set of points covered four times.

By adding a third dimension, we can display the same phenomena more clearly. If the patrons on the rim of the stadium shine their laser lights not straight ahead but up at a 45-degree angle, the rays will form a pattern in the space above the stadium. From the viewpoint of a blimp high above the stadium, the pattern looks the same, but from the side we see a far more intriguing configuration. The rays intersect one another to form lines of double points, and they interfere with each other to form bright curves of cusps. At certain points, arcs of double points come together with arcs of cusps to form even more complicated singularities. Strips of light radiating outward from positions spaced along the rim of the stadium interlace like the fingers of two hands coming together. If we send up lights from all points of the rim, they form a surface of rays, known in geometrical optics as "the catastrophe surface of the normal mapping."

This surface with its curves of singularities holds the history of all the parallel curves of the original curve. We can obtain the different parallel curves of the ellipse by slicing this surface with horizontal planes. A curve at one level is just the corresponding

Thick bands of light sent out from the ellipse at a 45-degree angle from its plane produce the catastrophe surface of the ellipse in space. Starting from the left, this figure is viewed from above, then from an angle, then filled in.

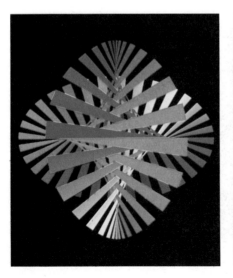
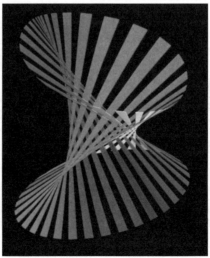
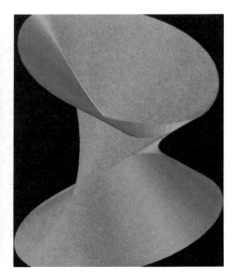

parallel curve lifted out of the plane a distance equal to its planar distance to the original ellipse. When the horizontal slice goes through one of the two arcs of double points, the corresponding parallel curve has a pair of double points. When it goes through a cuspoidal curve on this surface, the corresponding parallel curve will have a cusp at that point.

We can relate this construction back to the three-dimensional Laguerre geometry of circles in the plane, mentioned earlier in this chapter. In that geometry, a point in three-space corresponds to the circle centered at the point in the horizontal plane specified by the first two coordinates, with its radius given by the third coordinate. The points of a 45-degree line perpendicular to a horizontal line at a point correspond to circles in the plane that are tangent to the line at the point. Therefore, the collection of these 45-degree lines forming the catastrophe surface of a curve corresponds to the collection of circles that are tangent to the curve at at least one point. A great deal of the geometry of the curve is contained in this surface. For example, the double point curve, where the catastrophe surface intersects itself, corresponds to the collection of circles that are tangent to the curve at two or more points.

We can return to gain a new insight into previously understood cases by deforming the ellipse back into a circle. The rays moving up at a 45-degree angle will move along during the deformation, until the entire focal behavior spread out in the case of the ellipse coalesces to a single point, the origin of a double cone. We can think of the focal curve of the ellipse as a perturbation of this highly singular conical point. This picture of a conical point will be useful to us later on in our investigations of surfaces of revolution.

Wave Fronts in Three-Dimensional Space

Mathematicians have discovered many properties of plane curves by relating them to the geometry of the catastrophe surface in three-dimensional space. That process of taking the life history of a changing object and stretching it out in a different spatial direction makes it possible for us to translate temporal phenomena in one space into static configurations in another space. By applying all of our visualization techniques to the static catastrophe surface, we can study the phenomena in new ways.

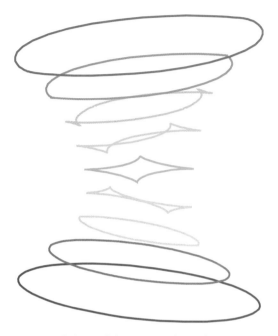

Horizontal slices of the catastrophe surface give parallel curves of the ellipse displaced in space.

A circle's catastrophe surface is a double cone.

This cutaway view reveals the spherical waves emanating from a point in space.

We could now go on to examine other plane curves by the same methods, investigating relationships between a curve and its focal curve to understand better how a planar object radiates waves. But since our primary object is to explore different dimensions, we will now take the experience we have gained in studying objects in two dimensions and apply it to phenomena in three and more dimensions.

What happens when our surrounding space is not two-dimensional but three-dimensional? When the simplest object, a point, radiates heat or sound or light, then the waves that emanate through the surrounding space are concentric spheres. The precise speed of the waves depends on the system's physical characteristics, but the geometric shape of the waves will invariably be described by these spherical wave fronts.

We could portray the entire history of the waves from a point by drawing a collection of concentric spheres, but the larger spheres will totally obscure the others. One remedy would be to use transparent spheres, or we could make use of the symmetry of the wave front by showing not the entire wave front but rather half of it, a lower hemisphere. The successive waves from a point will then be nested hemispherical shells, which we can see all at once. This representation clearly shows that the concentric circles in the plane of the equator display precisely the history of the waves emanating from a point in that plane.

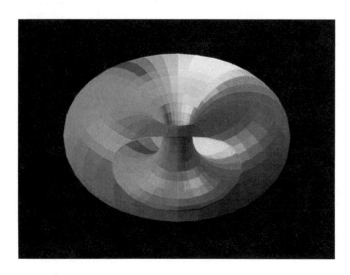

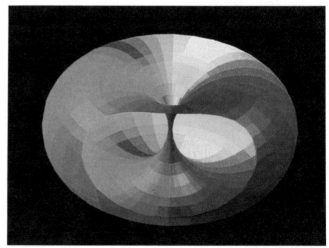

The simplest one-dimensional object, a straight line, sends out wave fronts in space that are circular cylinders, all with the same axis. A cross section perpendicular to the axis will reproduce the same concentric circle pattern generated by a point in the plane. If we slice the cylinders by a plane containing the original line, we obtain the lower-dimensional case of a line radiating pairs of parallel lines.

Once again, the simplest closed curve is a circle. The waves that emanate from a circle in space are surfaces of revolution. At the beginning, such a wave front will be a torus, the surface formed by revolving a small circle centered at a point of the original circle and lying in a plane perpendicular to the plane of that circle. As the wave front moves outward from the circle, it begins to run into itself and develop singularities.

To prevent the larger parallel surfaces from obscuring what is happening inside, we can use slicing techniques to expose the family of parallel surfaces. If we slice by the plane of the original circle, we obtain a family of pairs of concentric circles identical to the wave fronts from a circle in the plane.

However, if we slice by a plane perpendicular to the plane of the circle and passing through the center of the circle, we obtain something quite new. Such a plane intersects the original circle in a pair of opposite points situated symmetrically with respect to a line through the center of the original circle. From each of these

Cutaway views show that waves emanating from a circle in space form torus surfaces, which eventually intersect themselves to form horn cyclides and spindle cyclides.

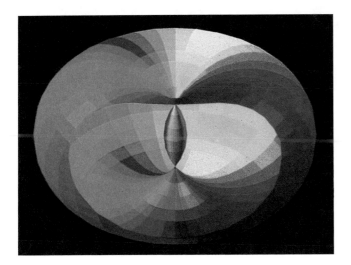

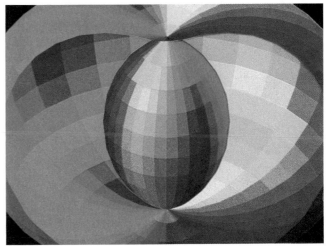

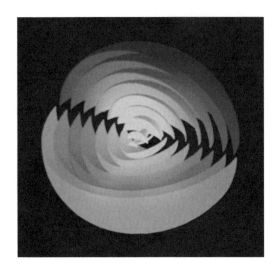

The parallel surfaces of an ellipsoid of revolution.

Left: Parallel surfaces of an ellipsoid with unequal axes. *Right:* The focal curves of the ellipsoid with unequal axes are traced out by cuspidal edges of parallel surfaces.

points there proceeds a family of concentric circles. As the wave front moves inward, the two circles do not intersect at first, but eventually the two outermost circles touch at one point and subsequently meet at a pair of points on the symmetry line.

The parallel surface at each stage is obtained by forming the surface of revolution of the pair of curves about the symmetry axis. At first this surface is a torus without singularities. When the circular waves first touch, their surface of revolution is what nineteenth-century geometers called a "horn cyclide." When the circular waves intersect, then their surface of revolution becomes a "spindle cyclide." It has two singular points, each resembling the singular point of a double cone.

We may also consider the parallel surfaces radiating from a surface in three-dimensional space. As before, if the distance to a smooth surface is small enough, the parallel surfaces will themselves be smooth. As the distance increases, the parallel surfaces may develop singularities. For example, the parallel surfaces to a sphere will collapse down to the sphere's center and reemerge, so there is just one focal point.

For an ellipsoid of revolution, formed by revolving an ellipse around one of its axes of symmetry, the parallel surfaces will themselves be surfaces formed by revolving the parallel curves of the ellipse around the same axis. The singularities of these parallel surfaces will be circular curves of cusps and "spindle points"

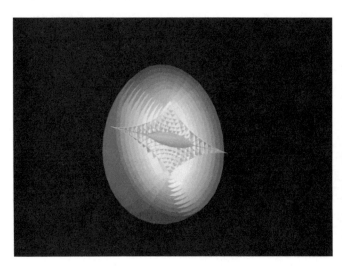

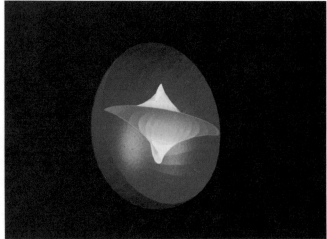

and circles of double points. These singularities all lie either on the surface of revolution of the evolute curve of the ellipse or on a segment along the axis of revolution.

But what happens as we deform an ellipsoid of revolution to an ellipsoid with three unequal axes? This question was raised over 130 years ago by the British mathematician Arthur Cayley. By a great effort he was able to construct one picture to show the form of the focal surfaces for a single example. Today, using computer graphics, we can create entire families of parallel surfaces and their associated focal surfaces.

8 COORDINATE GEOMETRY

Underlying everything we have done with dimensions is a basic framework called coordinate geometry or analytic geometry. Again and again we have encountered strings of numbers describing coordinates of a location or a shape. The identification of a point in space with a sequence of numbers is the basic connection between geometry and algebra. The fundamental relationships among points in the plane are mirrored in the relationships among pairs of numbers, while triples of numbers mirror the relationships among points in space. Geometric transformations like scaling and projecting correspond to transformations of the coordinate pairs or triples. Facts about geometry are translated into algebraic facts, and vice versa. The mathematics dealing with these transformations is called linear algebra.

Unfortunately, this effective way of dealing with the mathematics of dimensions has also served to isolate many of its most beautiful results from a general readership. In this book, I have purposefully chosen to treat geometric subjects from what is called a synthetic point of view, using coordinate representation sparingly and not developing the algebraic aspects extensively.

The synthetic viewpoint dominated geometric studies from the time of the Egyptians and Greeks until the seventeenth century, when the invention of analytic geometry by René Descartes

The intricate linear paintings of James Billmyer connect points at particular locations on the picture plane, leading the viewer off the page and back again in four different directions corresponding to different colors.

set the stage for the development of higher-dimensional coordinate geometry two centuries later. At first, mathematicians restricted their application of analytic geometry to numbers on the number line and to number pairs in the plane, but by the beginning of the nineteenth century, it was well understood that the algebra that worked to describe the number line and the coordinate plane also extended to three-dimensional space.

Today we treat such progressions very naturally. A theorem about objects in the plane, when expressed in coordinate form, often suggests a corresponding theorem in space—instead of writing two coordinates, we simply write three. But if we write two or three, why not four? As the algebra is practically the same, the theorems about number pairs and number triples extend to yield formal theorems about manipulations of quadruples of numbers. In analytic geometry, the most powerful results occur when we express some geometric relationship in coordinate form, then manipulate the number pairs or triples algebraically, and finally reinterpret the effect of these transformations on the original points in the plane or in space. But what geometric interpretation can we give for the analogous manipulations of number quadruples? And what happens when we try to interpret abstract relationships among sequences of 5 or 11 or 26 coordinates?

For the most part, mathematicians concerned with higher dimensions have been content to make use of the formal statements of linear algebra, retaining the geometric vocabulary but abandoning the attempt to visualize the concepts in concrete terms. All this is beginning to change with the advent of modern graphics computers, which literally do not know what dimension they are in. If we enter a collection of number pairs, the computer will display them as points on a television screen. If we enter number triples, the computer will first of all replace each triple by a pair according to some rule, then display the point pairs. The methods used to determine the screen coordinates of a point come from linear algebra.

Coordinates and Axes

In order to set up the framework that will enable us to deal easily with objects in higher-dimensional space, we develop briefly some of the structures we can use to locate numbers, number pairs, and number triples, in coordinate geometries of one, two, and three dimensions.

Coordinates on the number line.

On a line, we may choose an origin, labeled 0, and a unit point, labeled 1. The origin sets a starting point, and the distance between the origin and the unit point establishes a scale. Any point on the ray from 0 through 1 is specified by its distance from the origin, represented as a multiple of the distance from 0 to 1. Points on the oppositely directed ray are also identified by their distance, but in this case the number is preceded by a negative sign. In this way, each point corresponds to a real number, its coordinate, and each real number corresponds to a single point on the line. The line is a one-dimensional space.

The plane is a two-dimensional space. In order to set up coordinates for points of the plane, we start with two number lines, called coordinate axes, intersecting at their common origin, labeled (0, 0). We label points on the first axis $(x, 0)$, where x is the coordinate on the first number line, and points on the second axis $(0, y)$. Through any point in the plane we may draw lines parallel to the coordinate axes, meeting the first axis in a point labeled $(x, 0)$ and the second in a point labeled $(0, y)$. The point's location with respect to this choice of coordinate axes is then completely determined by the number pair (x, y). The point (x, y) is the fourth

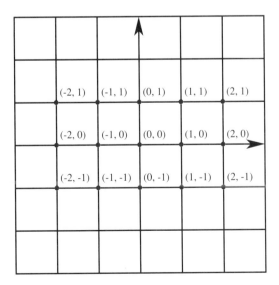

Coordinates in the plane.

Weavings designed by Joan Erikson and executed by Mary Schoenbrun use a two-dimensional gridwork to illustrate the interplay between different stages of the life cycle, as described in the theories of personality development of Erik and Joan Erikson. The warp and the weave make equal contributions in the first example, and, in the second, the height of each rectangle indicates the duration of the stage.

vertex of a parallelogram with one vertex at the origin and the other vertices at $(x, 0)$ and $(0, y)$. In coordinate geometry, we express this geometric construction by defining the coordinates of the sum of two point pairs as the sum of the corresponding coordinates:

$$(a, c) + (b, d) = (a + b, c + d)$$

For an arbitrary point, we find $(x, y) = (x, 0) + (0, y)$, so any point can be expressed as a sum of points on the coordinate axes. In this coordinate system, we can define the unit square with the four corner points labeled $(0, 0)$, $(1, 0)$, $(1, 1)$, and $(0, 1)$.

In order to set up coordinates for points in ordinary three-dimensional space, we start with three number lines intersecting at their common origin $(0, 0, 0)$. We label points on the first axis $(x, 0, 0)$, points on the second axis $(0, y, 0)$, and points on the third $(0, 0, z)$. As in the case of the plane, we define the coordinates of the sum of two triples as the sum of the corresponding coordinates. The sums of points on the first two axes determine the 1-2-coordinate plane containing all points of the form $(x, y, 0) = (x, 0, 0) + (0, y, 0)$. Similarly, the points on the 1-3-coordinate plane have coordinates $(x, 0, z)$, and the points on the 2-3-coordi-

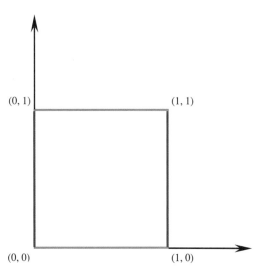

The unit square in two-dimensional coordinate space.

nate plane are of the form $(0, y, z)$. Through any point in three-space labeled (x, y, z), there are three planes parallel to these coordinate planes, meeting the first axis in the point $(x, 0, 0)$, the second in the point $(0, y, 0)$, and the third in a point $(0, 0, z)$. Any point (x, y, z) can therefore be written as a sum of points on the three coordinate axes. The point (x, y, z) is the eighth vertex of a parallelepiped with one vertex at the origin. In this coordinate system, we can define the unit cube with the eight corner points labeled $(0, 0, 0)$, $(1, 0, 0)$, $(1, 1, 0)$, $(0, 1, 0)$, $(0, 0, 1)$, $(1, 0, 1)$, $(1, 1, 1)$, and $(0, 1, 1)$.

It is no accident that we have used the same basic language of axes and coordinates in describing the two-dimensional geometry of the plane and the three-dimensional geometry of ordinary space. Such a common description expresses a profound relationship between the two spaces. As we describe objects in the plane, we can often easily identify corresponding objects in space. And the analogy goes further—we may use the same language to define a system of number quadruples, a four-dimensional hyperspace.

In the case of four-dimensional space, we can no longer base our concepts on familiar constructions of lines and planes. We begin with the coordinate representation itself, and ask what we can say about the space of all four-tuples of real numbers (x, y, u, v). Guided by our experience in the plane and in three-space, we designate the points of the form $(x, 0, 0, 0)$ to be the first coordinate axis. Similarly, the points of the form $(0, y, 0, 0)$ determine the second coordinate axis, while points of the form $(0, 0, u, 0)$ compose the third axis, and those of the form $(0, 0, 0, v)$ constitute the fourth coordinate axis. The point with label $(0, 0, 0, 0)$, the intersection of these four axes, is the origin of the coordinate system. As in the case of the plane and three-space, we add two four-tuples by adding their corresponding coordinates. Thus every point in four-space can be expressed as the sum of four points lying on the coordinate axes.

Any pair of these coordinate axes determines a coordinate plane, for example the 2-3-coordinate plane containing all points of the form $(0, y, u, 0)$. Any three of the four coordinate axes determine a coordinate hyperplane. Through any point in four-space there are four hyperplanes parallel to the coordinate hyperplanes, meeting the first axis in $(x, 0, 0, 0)$, the second in $(0, y, 0, 0)$, the third in $(0, 0, u, 0)$, and the fourth in $(0, 0, 0, v)$. With respect to this choice of coordinate axes, any point is completely determined by the number quadruple (x, y, u, v), and the point is the

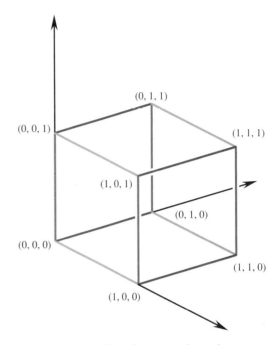

The unit cube in three-dimensional coordinate space.

The unit hypercube in four-dimensional coordinate space.

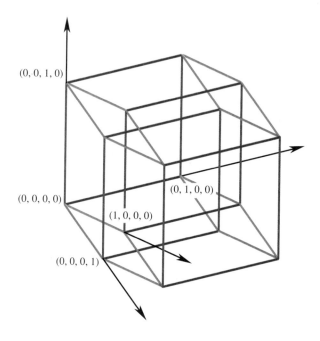

$(0, 0, 1, 0)$

$(0, 0, 0, 0)$

$(0, 1, 0, 0)$

$(1, 0, 0, 0)$

$(0, 0, 0, 1)$

sixteenth vertex of a parallelotope with one vertex at the origin. In this way we obtain a coordinate system for four-dimensional space. In this coordinate system, we can define the unit hypercube with the 16 corner points labeled:

$$(0, 0, 0, 0), (1, 0, 0, 0), (1, 1, 0, 0), (0, 1, 0, 0),$$
$$(0, 0, 1, 0), (1, 0, 1, 0), (1, 1, 1, 0), (0, 1, 1, 0),$$
$$(0, 0, 1, 1), (1, 0, 1, 1), (1, 1, 1, 1), (0, 1, 1, 1),$$
$$(0, 0, 0, 1), (1, 0, 0, 1), (1, 1, 0, 1), (0, 1, 0, 1)$$

There is nothing to stop us from carrying out the same abstract description of a collection of five-tuples in five-space, or n-tuples in n-space. In one sense, we can say that n-dimensional space is exactly the collection of all n-tuples of real numbers, but this ignores the rich geometric content of n-dimensional geometry. We can gain additional insight into the relationships of number pairs, triples, or n-tuples by graphing them in lower-dimensional spaces, like the plane or three-space.

Lengths and the Generalized Pythagorean Theorem

One of the greatest advantages of analytic geometry is that in a coordinate system of any dimension there is an explicit formula for the distance between two points, found by generalizing the Pythagorean theorem.

On a one-dimensional coordinate system, we may determine the distance between two points by finding the absolute value of the difference of their coordinates. The distance between the points labeled a and b is then $|a - b|$.

In the plane, any two points (a, c) and (b, d) may be joined by a segment, and this segment is a diagonal of a unique rectangle with edges parallel to the coordinate axes. Because the base of this rectangle has length $|a - b|$ and because the height of the rectangle is $|c - d|$, the Pythagorean theorem tells us that the length of the diagonal is given by $\sqrt{(a - b)^2 + (c - d)^2}$. For example, the diagonal of the unit square is the length of the line from $(0, 0)$ to $(1, 1)$, or $\sqrt{2}$.

The Pythagorean theorem can be extended to three-dimensional space by applying it to the length of the longest diagonal of a rectangular box. This diagonal is the hypotenuse of a right triangle having one side on the edge of the box and the other along a diagonal of a rectangular face of the box. We may apply the Pythagorean theorem once to get the length of the diagonal of the face and again to get the long diagonal of the box. The resulting formula for three-space is a direct generalization of the formula in the plane. Instead of taking the square root of the sum of the

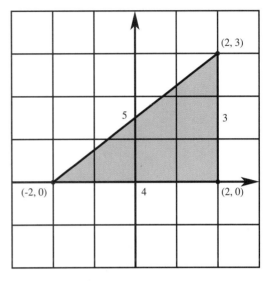

Applying the distance formula in the plane.

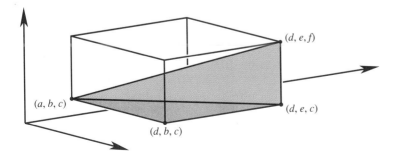

To find the distance formula in three-space, we apply the planar distance formula twice, once to find the length of the diagonal of one of the faces, and then to find the hypotenuse of a right triangle having this diagonal and an edge of the cube as its sides.

squares of the two sides of a rectangle, we take the square root of the sum of the squares of the three sides of the rectangular box. In analytic geometry terms, this means that the distance between the point with coordinates (a, b, c) and another labeled (d, e, f) is the square root of the sum of the squares of the differences of the corresponding coordinates, or $\sqrt{(a-d)^2 + (b-e)^2 + (c-f)^2}$. For example, the length of the longest diagonal of the unit cube in three-space is the distance from $(0, 0, 0)$ to $(1, 1, 1)$, or $\sqrt{3}$.

The generalization of the distance formula to higher dimensions is straightforward. By applying the Pythagorean theorem to a succession of planar triangles with sides given by edges or diagonals of the hypercube, the distance formula expresses the distance between two points as the square root of the sum of the squares of the differences of the coordinates. Thus the length of the longest diagonal of a unit hypercube is the distance between $(0, 0, 0, 0)$ and $(1, 1, 1, 1)$, namely $\sqrt{4} = 2$.

Coordinates for the n-Simplex

We have already seen how the vertices of an n-cube can be represented in n-space using only zeros and ones as coordinates. It is often somewhat harder to give a simple coordinate description of an n-simplex in n-space. It is not always a difficult task—there is a perfectly good choice for the vertices of a three-simplex in three-space obtainable by taking four of the eight vertices of a three-cube, for example $(0, 0, 0)$, $(1, 1, 0)$, $(0, 1, 1)$, and $(1, 0, 1)$. We know that this tetrahedron is regular because the distance between any two of its vertices is $\sqrt{2}$. The remaining four vertices of the cube, $(1, 0, 0)$, $(0, 1, 0)$, $(0, 0, 1)$, and $(1, 1, 1)$, also determine a regular tetrahedron. These two overlapping tetrahedra fit together to form a figure called the *stella octangula*, and their intersection is the dual octahedron contained in the cube.

However, if we want to find coordinates for the two-simplex in the plane, the situation is more complicated. If we choose the two coordinate pairs to be $(0, 0)$ and $(1, 0)$, then the third coordinate will be either $(1/2, \sqrt{3}/2)$ or $(1/2, -\sqrt{3}/2)$, and so we obtain fractions and irrational numbers as coordinates. There is no way to avoid using irrational numbers in the coordinates of an equilateral triangle as long as we stay in the plane. If, however, we are willing to go to three-space, then we can find a perfectly simple

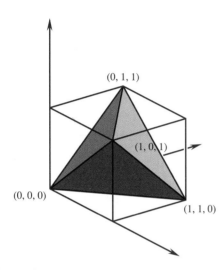

The regular tetrahedron in the unit cube.

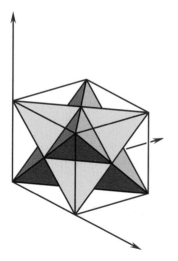

The stella octangula formed by two intersecting tetrahedra in the unit cube.

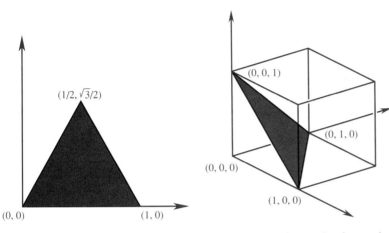

Irrational coordinates for the regular two-simplex in the plane.

Rational coordinates for the regular two-simplex in three-space.

collection of points forming the vertices of a regular two-simplex, for example three of the vertices of the second regular tetrahedron given above, $(1, 0, 0)$, $(0, 1, 0)$, and $(0, 0, 1)$.

Not only do the points with one coordinate 1 and all the rest 0 give a satisfactory solution to the problem of representing a two-simplex with simple coordinates, they also provide a method for finding coordinates of a simplex of any dimension. To find a coordinate representation for the $n + 1$ vertices of an n-simplex, we can take the points at unit distance along the coordinate axes in $(n + 1)$-space. This is a natural approach from the point of view of slicing since the configuration of vertices near each corner of an $(n + 1)$-cube is an n-simplex. For example, when we slice perpendicular to the long diagonal of a hypercube from $(0, 0, 0, 0)$ to $(1, 1, 1, 1)$, one of the slicing hyperplanes contains the four vertices $(1, 0, 0, 0)$, $(0, 1, 0, 0)$, $(0, 0, 1, 0)$, and $(0, 0, 0, 1)$. Since the distance between any two of these vertices is $\sqrt{2}$, they are the vertices of a regular three-simplex.

Coordinates for Hypercube Slices

When we analyzed objects in three- and four-space from a synthetic viewpoint, we learned quite a bit about the structure of cubes and hypercubes by keeping track of a sequence of slices perpendicular to the longest diagonal. The coordinate approach gives us a fresh look at this procedure. Some of the important geometric relationships among the slices already appear in the patterns of the coordinates.

As we slice a unit cube by planes perpendicular to the long diagonal, the first slice we get is the single vertex $(0, 0, 0)$. Then comes the equilateral triangle with vertices $(1, 0, 0)$, $(0, 1, 0)$, and $(0, 0, 1)$, having coordinates adding up to 1. As the slicing plane moves farther from the origin, the next vertices we encounter are those having coordinates adding up to 2—$(0, 1, 1)$, $(1, 0, 1)$, and $(1, 1, 0)$—also forming the vertices of an equilateral triangle. The final vertex that we hit is $(1, 1, 1)$, the unique vertex of the cube having coordinates adding up to 3.

We may carry out the same analysis for the hypercube in four-space. As we slice the hypercube perpendicular to the long diagonal from $(0, 0, 0, 0)$ to $(1, 1, 1, 1)$, the first slice is again the origin, $(0, 0, 0, 0)$. Next comes the three-simplex with vertices $(1, 0, 0, 0)$, $(0, 1, 0, 0)$, $(0, 0, 1, 0)$, and $(0, 0, 0, 1)$, having coordinates adding up to 1. In general, the vertices lying in a particular hyperplane perpendicular to this long diagonal will have coordinates adding

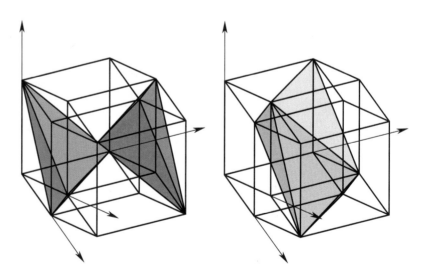

Hyperplane slices of the hypercube containing vertices with coordinates adding to 1 or 3 are tetrahedra (left), while the slice having vertices with coordinates adding to 2 is an octahedron (right).

up to a given number from 0 through 4. The unique vertex with coordinates adding to 0 is the origin, and the opposite vertex $(1, 1, 1, 1)$ is the unique vertex of the hypercube having coordinates adding up to 4. The slice containing vertices with coordinates adding up to 3 will be a three-simplex having vertices $(0, 1, 1, 1)$, $(1, 0, 1, 1)$, $(1, 1, 0, 1)$, and $(1, 1, 1, 0)$. Exactly in the middle we find a very interesting slice, where the coordinates add up to 2. There are six vertices in this slice—$(1, 1, 0, 0)$, $(1, 0, 1, 0)$, $(1, 0, 0, 1)$, $(0, 1, 1, 0)$, $(0, 1, 0, 1)$, and $(0, 0, 1, 1)$—and these form the vertices of a regular octahedron halfway through the hypercube.

Note that we find six vertices in the middle level of the hypercube since there are six ways to choose two zeros and two ones for the four coordinates. By studying the combinations of zeros and ones, we can predict the number of vertices in each slice. For the n-cube with vertices having coordinates 0 or 1, the slice perpendicular to the longest diagonal will start with the vertex having all coordinates 0 and end with the vertex having all coordinates 1. The slice containing vertices having coordinates adding up to k will contain $C(n, k)$ vertices, the number of ways to choose k coordinates equal to 1 and the rest 0. As this observation suggests, the numbers of vertices in hyperplane slices appear as the numbers in the rows of Pascal's triangle.

Coordinates for Regular Polyhedra

We have just obtained a set of coordinates for the vertices of a regular three-dimensional octahedron thought of as the middle slice of a hypercube in four-dimensional space. It is also easy to give a three-dimensional coordinate description for the octahedron by taking advantage of the fact that the octahedron is the dual of the cube: the vertices of a regular octahedron can be obtained as the centers of the six square faces of a cube. If we choose coordinates for the vertices of the cube, we can figure out the coordinates of the centers of square faces and obtain coordinates for the vertices of the octahedron. In the previous paragraph, we studied a cube with vertices having all coordinates 0 or 1, but in considering the dual, it turns out to be more convenient to start with a cube centered at the origin having all coordinates -1 or 1. The square faces of this cube are given by fixing one of the coordi-

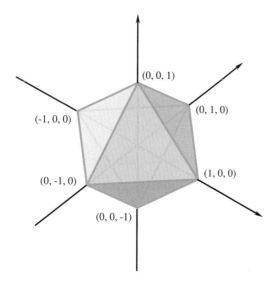

Coordinates for the octahedron.

nates. For example, the square face where all vertices have first coordinate -1 is given by the four vertices $(-1, 1, 1)$, $(-1, -1, 1)$, $(-1, -1, -1)$, and $(-1, 1, -1)$, and the center of this square is $(-1, 0, 0)$. In general, the centers of the squares are the six points at unit distance on the positive and negative coordinate axes, $(\pm 1, 0, 0)$, $(0, \pm 1, 0)$, and $(0, 0, \pm 1)$. These six points are the vertices of a regular octahedron with side length $\sqrt{2}$. In n-dimensional space, this same construction yields the coordinates of the $2n$ vertices of the n-dimensional cube-dual as the points at unit distance from the origin on the positive and negative coordinate axes.

What about the other regular polyhedra in three-space? We can exploit the symmetry of the icosahedron to come up with some fairly satisfactory coordinates, involving just one irrational number. And a very important number it is. We start by observing that for every edge of the icosahedron there is a parallel edge on the opposite side. We may situate the icosahedron in a cubical box so that all 12 vertices are on the boundary of the box and in fact so that the intersection with the boundary of the box consists of six edges parallel to the coordinate axes. As provisional coordinates for these segments, we may choose $(\pm 1, 0, \pm t)$, $(0, \pm t, \pm 1)$, and $(\pm t, \pm 1, 0)$, where t is a number to be chosen later. In general, the polyhedron with these 12 vertices will have edges of two lengths: $2t$ for the edges on the boundary of the box and $\sqrt{1 + t^2 + (1 - t)^2}$

Choosing different values of t opens up the octahedron ($t = .1$) by inserting isosceles triangles along its edges (left). For a special value ($t \approx .618$), the triangles are all equilateral and we have coordinates for the regular icosahedron (right).

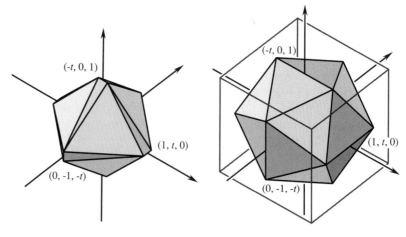

for the other edges. In order for the icosahedron to be regular, the lengths of all of these segments should be equal. This condition leads to the algebraic equation $t^2 + t - 1 = 0$, with positive solution $t = (-1 + \sqrt{5})/2$. This important number appears in all sorts of contexts involving the concepts of growth and form. It is called the *golden ratio*, and it expresses the ratio of a side of a regular pentagon to one of its diagonals. We should not be surprised to see it appear in conjunction with the regular icosahedron since the vertex configuration of each corner of this object is a regular pentagon.

It is possible to get a set of coordinates for the regular dodecahedron by taking the centers of the faces of the icosahedron given above, but we can find the coordinates more directly by exploiting another relationship with the cube. We start with a particular diagonal of a pentagonal face of the dodecahedron, then choose diagonals in the adjacent faces so that the three diagonals meeting at a vertex are mutually perpendicular and all of the same length. If we continue this procedure, we obtain 12 diagonals fitting together to form the edges of a cube inscribed in the dodecahedron, with one edge for each of the 12 faces of the dodecahedron. By starting with different diagonals of the original pentagon, we end up with five different cubes; in this collection of cubes, each of the 60 diagonals of the dodecahedron is used exactly once. By choosing the 8 vertices of the cube first, we may use the symmetry of

The pattern of seeds in a sunflower head appears to form systems of spirals that obey precise rules of growth and form. Running clockwise are 55 spirals, and running counterclockwise are 34. The ratio of these is a close approximation of the ratio between a side of a pentagon and one of its diagonals—the golden ratio.

A regular dodecahedron contains five cubes along the diagonals of its faces.

this figure to find coordinates for the other 12 vertices of the dodecahedron. If the vertices of the cube are $(\pm 1, \pm 1, \pm 1)$, then the others are of the form $(\pm t, 0, \pm 1/t)$, $(0, \pm 1/t, \pm t)$, and $(\pm 1/t, \pm t, 0)$, where t is the same number appearing in the coordinates for the icosahedron.

Coordinates for Regular Polytopes

In the course of finding coordinates for the regular polyhedra, we have found coordinates for each of the three regular polytopes in n-dimensional space, namely the n-simplex, the n-cube, and the dual of the n-cube. For dimension n greater than four, these are the only possible regular figures, but in four-dimensional space there are three more regular polytopes, with 24, 120, or 600 cells. To prove that these polytopes actually exist, we can exhibit coordinates for their vertices.

The self-dual 24-cell, with octahedral faces, turns out to be the easiest to describe by coordinates, using a procedure similar to that for finding the dual. Instead of choosing points in the centers of highest-dimensional faces, we can choose points in the centers of edges or squares. This procedure always will yield a figure with a large amount of symmetry, but usually there will be cells of different shape. For example, in the three-dimensional cube, the 12 points found in the centers of edges are vertices of a cuboctahedron with 8 triangular faces and 6 square faces. The surprise is that the centers of the 24 squares in a hypercube turn out to be the vertices of a regular polytope in four-space, the self-dual 24-cell. If the coordinates of the vertices of a hypercube are all either 1 or -1, then a square in this hypercube is determined by the four vertices each having two coordinates fixed and the other two either 1 or -1. For example, one such square has vertices $(\pm 1, \pm 1, 1, 1)$, and its midpoint will be $(0, 0, 1, 1)$. The coordinates of the midpoints of all of the squares in the hypercube can be listed in six groups: $(\pm 1, \pm 1, 0, 0)$, $(\pm 1, 0, \pm 1, 0)$, $(\pm 1, 0, 0, \pm 1)$, $(0, 0, \pm 1, \pm 1)$, $(0, \pm 1, 0, \pm 1)$, and $(0, \pm 1, \pm 1, 0)$. We obtain eight of the octahedra in the 24-cell by fixing one coordinate at 1 or -1, for example $(\pm 1, 0, 0, 1)$, $(0, \pm 1, 0, 1)$, and $(0, 0, \pm 1, 1)$. The other 16 octahedra are obtained by choosing a vertex of the hypercube and replacing two of its coordinates by 0 in six different ways; for example, $(1, -1, -1, 1)$ yields the verti-

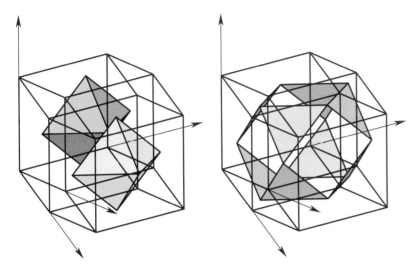

Slices of the centered 24-cell with final coordinate 1 or −1 are octahedra (left), while the slice with final coordinate 0 contains 12 vertices forming the vertices of a cuboctahedron (right).

ces $(1, -1, 0, 0)$, $(1, 0, -1, 0)$, $(1, 0, 0, 1)$, $(0, 0, -1, 1)$, $(0, -1, 0, 1)$, and $(0, -1, -1, 0)$.

By carefully exploiting their symmetry, it is possible to choose coordinates for all 600 vertices of the 120-cell and for all 120 vertices of the 600-cell. We can instruct a graphics computer to display the image of any of these polytopes in the plane, or we can use the animation capabilities of the machine to create a motion picture of the image in three-space. There is a real art to choosing coordinates in a way that transmits the maximum amount of visual information. Some of the mathematical principles underlying this art are contained in the several books of the Canadian geometer H. S. M. Coxeter, one of the foremost developers of the theory of polyhedra and polytopes. He wrote his first paper in the subject in 1923 at the age of 16.

Complex Numbers as Two-Dimensional Numbers

Up to this point, we have considered pairs, triples, and n-tuples of real numbers, starting with the points on a one-dimensional real number line. Although the real numbers are sufficient for very many purposes in algebra and geometry, they are inadequate for solving some simple equations. For example, since the square of any real number is never negative, it is impossible to find a real number that solves the equation $x^2 = -1$. To construct a larger number system where this equation can be solved, mathematicians introduced a new symbol, i, with the property that $i^2 = -1$.

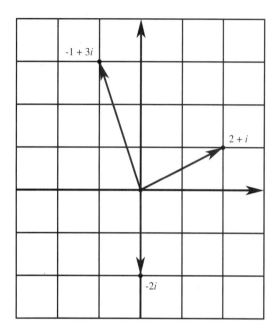

-1 + 3i

2 + i

-2i

Complex numbers in the coordinate plane.

In order still to be able to add numbers and multiply them by real numbers, they had to include in their system all numbers of the form $x + yi$, with x and y real. They called this collection the *complex numbers.*

Each complex number $x + yi$ corresponds to a number pair (x, y) in the plane, so we may say that the complex numbers form a two-dimensional collection. The two coordinates of the pair (x, y) are called the real part and the imaginary part of the complex number. We already know how to add number pairs and multiply them by real scalars, and this leads to rules for addition and scalar multiplication for complex numbers. The sum of $x + yi$ and $u + vi$ is $(x + u) + (y + v)i$, and the product of $x + yi$ by the real number c is $cx + (cy)i$.

But it is also possible to define a multiplication of one complex number with another, giving as the product of two complex numbers

$$(x + yi)(u + vi) = (xu + yui + xvi + yvi^2) = (xu - yv) + (yu + xv)i$$

Thus the product of two complex numbers is another expression of the same form. For any $x + yi$ not equal to 0, there is another complex number $u + vi$ such that $(x + yi)(u + vi) = 1$, specifically $u + vi = x/(x^2 + y^2) - yi/(x^2 + y^2)$. As these examples illustrate, the complex numbers share very many of the properties of the real numbers. In a sense, we can say that the essence of complex numbers is that they are the collection of number pairs endowed with the usual rule for addition and an unusual rule for multiplication: $(x, y)(u, v) = (xu - yv, yu + xv)$. The justification for this rule is that so many of the desirable algebraic properties of real numbers still hold. For example, since $(x, y)(1, 0) = (x, y)$, the number pair $(1, 0)$ is a unit element for multiplication. But in this new system, we also have an element $(0, 1)$ such that $(0, 1)(0, 1) = (-1, 0)$. Thus an element exists whose square is the negative of the unit element, the key property of the complex numbers. The most significant thing about this construction is that we can relate the algebraic properties of complex numbers to the geometric properties of the plane. The interplay between the algebraic and the geometric accounts for the rich structure of the complex number system and for its surprisingly many applications in science and engineering.

One of the most effective techniques in ordinary analytic geometry is the graphing of functions of one real variable on a two-dimensional grid. In ordinary analytic geometry, the equation $u = x^2$ can be graphed in the plane by plotting all pairs of the form (x, x^2). The geometric form of this graph is a parabola, symmetric with respect to the vertical axis and passing through the origin. This representation gives a tremendous amount of insight into the symmetry of the function as well as the location of its minimum points.

What about complex functions of a complex number? We can still talk about the equation $w = z^2$, where now z and w represent complex numbers. We can analyze the function algebraically all right, but how can we graph it? A problem with dimensions precludes a simple graph on paper. A single complex number already requires two real numbers, its real and imaginary parts, so we have two real coordinates for z and two more for w. The total graph requires four real dimensions, two for the domain and two for the range, giving a two-dimensional surface in four-dimensional space. To study such an object from a geometric point of view, we can try all of the techniques we have used in investigating hypercubes and other figures in four-dimensional space.

Already in the last century, mathematicians created plaster models to show projections of such complex function graphs into three-space, but it was often difficult to imagine how these various views could all arise from the same object in four-space. The fulfillment of this approach only came much later, in our own time, with the development of interactive computer graphics. Many modern graphics computers can rotate filled-in surfaces in four-dimensional space almost as fast as in three-space, so we can see what happens to the images of the graph of a complex function as we rotate it in four-space and project it into our ordinary viewing space.

A basic example demonstrates how the computer handles the equations of complex functions. We have already discussed the parabola in the plane, which is the graph of the real squaring function $u = x^2$. For complex numbers, the relation $w = z^2$, where $z = x + yi$ and $w = u + vi$, can be expressed in terms of the real coordinates x, y, u, and v. The usual rules of algebra give

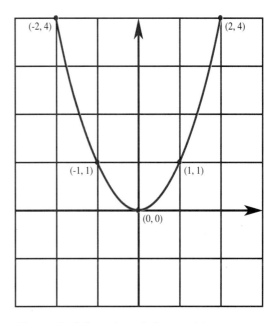

The graph of the real parabola in real two-space.

$$z^2 = (x + yi)^2 = x^2 + 2xyi + (yi)^2 = x^2 - y^2 + 2xyi = u + vi$$

Therefore $u = x^2 - y^2$ and $v = 2xy$. The analogue in four-space of the graph of the parabola in the plane is the collection of points of the form (z, z^2), but now each of these coordinates is determined by two real numbers, for a total of four real coordinates. We may list the two real and imaginary parts of z and the two real and imaginary parts of z^2 in a four-tuple $(x, y, x^2 - y^2, 2xy)$, and the collection of these four-tuples in four-dimensional space will be the graph of the complex squaring function.

Given any point in the (x, y)-plane, we may determine the other two coordinates and plot the points. As usual, one of the most effective ways of visualizing a collection of points in four-space is to project them into the plane or into three-space. The same instructions that enable us to create motion pictures of a rotating hypercube make it possible to project the graph of the complex squaring function from four-space. Projecting into the three-space of the first three coordinates yields $(x, y, x^2 - y^2)$, the graph of a real function of two variables, a hyperbolic paraboloid. Projecting into the hyperplane determined by the first, second, and fourth coordinates yields $(x, y, 2xy)$, a rotated hyperbolic paraboloid.

More interesting images appear when we project into the hyperplane of the last three coordinates. We obtain the graph of $(y, x^2 - y^2, 2xy)$, known as the imaginary part of the square root relation. The image is very different from the hyperbolic parabo-

Left: The graph of the real part of the complex parabola projected from four-space into three-space. *Right:* The graph of the imaginary part of the square root function.

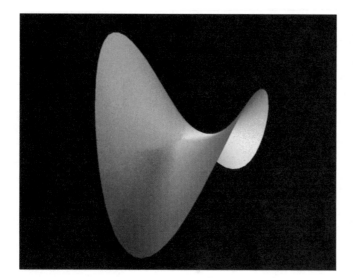
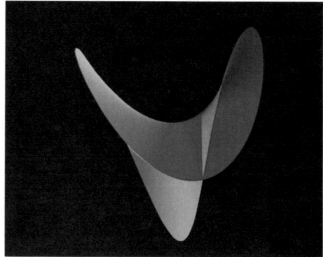

loids, possessing a singular point at the origin. The figure intersects itself along a ray of double points on a coordinate axis, ending in a so-called "pinch point" at the origin. This example provides one of the most difficult rendering challenges for a graphics computer and yields some of the most interesting images from the space of four dimensions.

Four-Dimensional Numbers: The Quaternions

When complex numbers were first invented, many people were suspicious. What real significance could such "imaginary" numbers have? But it was only a short time before mathematicians and scientists discovered a tremendous number of applications for these numbers. The complex numbers turned out to be precisely the system that best described patterns of flows in hydrodynamics and in electricity. These two-dimensional numbers have more than justified their existence.

One hundred years ago, Sir William Hamilton invented a number system of even higher dimension, the *quaternions*, a collection of four-dimensional numbers. The addition of quaternions is defined the same as addition of points in four-space, but the multiplication law is a complicated mixture of different formulas

Left: The graph of the real part of the cube root function ($w = z^3$) projected from four-space into three-space. *Right:* The graph of the imaginary part of the cube root function.

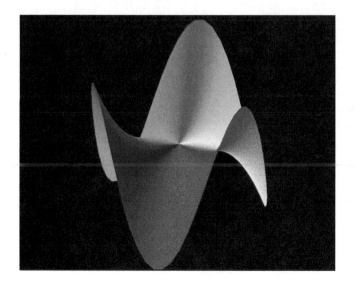
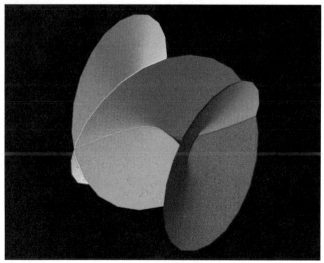

that come from vector calculus. Specifically we define $(a, b, c, d)(x, y, u, v) = (ax - by - cu - dv, ay + bx + cv - du, au - bv + cx + dy, av + bu - cy + dx)$. As in the case of the complex numbers, there is a unit element $(1, 0, 0, 0)$, and for every non-zero quaternion (a, b, c, d), there is a quaternion (x, y, u, v) such that $(a, b, c, d)(x, y, u, v) = (1, 0, 0, 0)$.

But would such numbers ever have practical applications? As it happens, these are exactly the numbers used to describe the orbits of the motion of pairs of pendulums. And quaternions have been recognized in recent years as one of the most effective means of communicating information about rotations to a graphics computer. Thus some of the most abstract algebraic constructions in four-space can be visualized as geometric phenomena.

Coordinates for Circles and Spheres

Up to this time, we have not made use of trigonometric expressions in describing geometric objects. In the spirit of coordinate geometry, these expressions are extremely useful for specifying points on circles, spheres, and hyperspheres. In the plane, a unit circle is the collection of points (x, y) at distance 1 from the origin. By the Pythagorean theorem, this condition may be expressed as the algebraic statement $x^2 + y^2 = 1$. Analogously, in three-space the unit sphere is defined as the collection of points at unit distance from the origin, so in terms of coordinates, the unit sphere is the collection of points (x, y, u) such that $x^2 + y^2 + u^2 = 1$. In four-space, the unit hypersphere is the collection of points (x, y, u, v) such that $x^2 + y^2 + u^2 + v^2 = 1$.

To determine coordinates for the unit circle, mathematicians invented two functions, cosine and sine. These functions describe the position on a circle of a point reached by starting at $(1, 0)$ and traveling counterclockwise around the circle a certain distance. If that distance is t, then the coordinates of the point on the unit circle are given by $(\cos(t), \sin(t))$.

From the coordinates for a circle, we can get geographical coordinates for the points of the unit sphere in three-space. Specifically, if t is the longitude of a point and s is its latitude, then the coordinates of the point are $(\cos(t)\cos(s), \sin(t)\cos(s), \sin(s))$. It is straightforward to show that the sum of the squares of these coordinates is 1, the defining property of the unit sphere.

Similarly, we may obtain points on the unit hypersphere by using three angular coordinates, t, s, r, and defining $(\cos(t)\cos(s)\cos(r), \sin(t)\cos(s)\cos(r), \sin(s)\cos(r), \sin(r))$. Another representation using angular coordinates t, s, and r is even more symmetric, namely $(\cos(t)\cos(s), \sin(t)\cos(s), \cos(r)\sin(s), \sin(r)\sin(s))$. In each case we can show that the sum of the squares of the coordinates is 1, thus establishing that the points lie on a unit hypersphere. In the second representation, for a fixed choice of the variable s, we obtain a circle of circles. Choosing s equal to 45 degrees, we obtain the Clifford torus, with the simple formula: $1/\sqrt{2}(\cos(t), \sin(t), \cos(r), \sin(r))$. In this way we define a collection of torus surfaces in the hypersphere, and these are precisely the surfaces that we have already seen in our study of stereographic projection and in our analysis of orbits of dynamical systems. The orbits for a pair of synchronized pendulums are given by $(\cos(t)\cos(s), \sin(t)\cos(s), \cos(t + c)\sin(s), \sin(t + c)\sin(s))$ for any fixed choice of s and c. It is remarkable that a graphics computer can take such simple equations and produce pictures of such rich quality.

Twisted bands link together to describe the orbits of a system of pendulums graphed in the sphere in four-space.

NON-EUCLIDEAN GEOMETRY and NONORIENTABLE SURFACES

In the middle part of the nineteenth century, mathematicians first realized that there were different kinds of geometries, geometric systems that did not obey Euclid's rules for plane and solid geometry. The idea that there could be geometries of higher dimensions was disturbing to those philosophers who subscribed to a "realistic" view of geometry, and they were even more upset by the suggestion that there could be different kinds of two-dimensional geometry. Certainly it was well known that the geometry on various surfaces, such as the sphere, was fundamentally different from the geometry of the flat plane. But did the geometry on part of a spherical surface constitute a different two-dimensional geometry, as some mathematicians now claimed? What did it mean to have different kinds of geometry? To explain their ideas, the mathematicians who proposed the new theories resorted to the time-honored device of analogy, and they even asked people to empathize with a two-dimensional creature confined to move along a curved surface. Visualizing dimensions was important in all phases of the controversial new geometries.

A glass sculpture presents a view in three-space of a famous nonorientable surface, the Klein bottle.

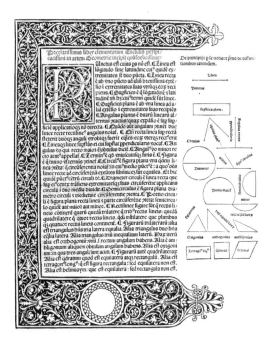

The first page of the first printed edition of Euclid's *Elements*, published in 1482.

The Axioms of Euclidean Plane Geometry

For well over two thousand years, people had believed that only one geometry was possible, and they had accepted the idea that this geometry described reality. One of the greatest Greek achievements was setting up rules for plane geometry. This system consisted of a collection of undefined terms like point and line, and five axioms from which all other properties could be deduced by a formal process of logic. Four of the axioms were so self-evident that it would be unthinkable to call any system a geometry unless it satisfied them:

1. A straight line may be drawn between any two points.
2. Any terminated straight line may be extended indefinitely.
3. A circle may be drawn with any given point as center and any given radius.
4. All right angles are equal.

But the fifth axiom was a different sort of statement:

5. If two straight lines in a plane are met by another line, and if the sum of the internal angles on one side is less than two right angles, then the straight lines will meet if extended sufficiently on the side on which the sum of the angles is less than two right angles.

   Because this axiom was much more complicated than the previous axioms, it seemed more like a theorem than a self-evident proposition. Since all attempts to deduce it from the first four axioms had failed, Euclid simply included it as an axiom because he knew he needed it. For example, some axiom like this one was necessary for proving one of Euclid's most famous theorems, that the sum of the angles of a triangle is 180 degrees. Mathematicians found alternate forms of the axiom that were easier to state, for example:

5′. For any given point not on a given line, there is exactly one line through the point that does not meet the given line.

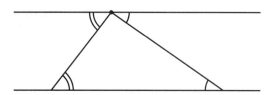

Once we can draw a unique line through one vertex of a triangle not meeting the line containing the opposite side, we can use alternate interior angles to see that the sum of the angles of a triangle is the same as a straight angle, 180 degrees.

This form of the fifth axiom became known as the parallel postulate. Although it was simpler to understand than Euclid's original formulation, it was no easier to deduce from the earlier axioms. The attempt to deduce the fifth axiom remained a great challenge right up to the nineteenth century, when it was proved that the fifth axiom did not follow from the first four.

The great advantage of expressing geometry as an axiomatic system was that it no longer was necessary to memorize long lists of independent facts about the nature of the universe—one only had to know a small set of axioms, and by applying to them the rules of inference, one could reconstruct the entire collection of geometric truths.

There was little doubt that the Greeks were attempting to describe a real world when they formulated their geometry, even though it might have been an ideal sort of world, realized only abstractly "in the mind of God." Many mathematicians, now as well as in the distant past, believe that the complete structure of mathematics is something that exists in itself and that it is only gradually discovered by human beings laboring to uncover its mysteries. Even though the framers of the early axiom systems would refer to point and line as undefined terms, they fairly clearly thought of them as real objects, and they thought that the system they were developing was a progressively more and more elaborate and accurate description of the real world. The progress of algebra, on the other hand, was not quite so settled, and people accepted changes in viewpoint there more readily than in the very traditional field of geometry.

Noncommutative Algebra

For the realists, especially the followers of the influential German philosopher Immanuel Kant, the essence of geometry was that it described experience. The suggestion that some new system of statements deserved to be called geometry was a threat. Yet the followers of Kant did not object when formulas in algebra no longer seemed to describe reality. Certainly a great many of the most familiar algebraic relationships originated from real problems, some of them geometric and some from economics and physical science. Parabolas were familiar as conic sections and as

Immanuel Kant was the chief philosopher who championed a realistic view of geometry.

the paths of projectiles, and the simple formula for a parabola could easily be graphed using the techniques of analytic geometry. Volume formulas suggested cubic equations, and they were also easy to graph and analyze. But it was not much harder to use the same techniques to analyze polynomials of degree four or five or higher. Few objected that this was not algebra, even though it no longer had a convenient geometric interpretation.

However, there was some resistance to algebraic innovation the first time someone suggested an algebraic operation that was not commutative. By this stage, mathematicians had written down axiom systems describing the ordinary algebra of real numbers. Among the rules describing addition and multiplication were two stating that the order of adding or multiplying was not important: the sum of a and b was the same as the sum of b and a, and the same was true for the product. Later, mathematicians began to realize that laws for combining elements in other sorts of systems satisfied most of the axioms of ordinary algebra, and so these systems behaved in large measure like numbers. An important example is the collection of symmetries of a square. We can rotate a square by one quarter turn in a counterclockwise direction to move all four vertices. Rotating the square again results in a half turn and rotating once again yields a three-quarter turn. It does not matter in which order we make one turn after another. The rotational symmetries of the square form a commutative system.

But the collection of all symmetries of the square is not a commutative system. We can reflect a square across its diagonal, keeping two vertices fixed and interchanging the other two, or reflect about a horizontal or a vertical line through the center, moving all four vertices. When we combine such reflections with rotations, the order in which we apply them makes a difference. If we reflect across a diagonal and then rotate by a quarter turn, the effect is the same as a vertical reflection. On the other hand, if we first rotate by a quarter turn and then reflect across the diagonal, the effect is the same as a horizontal reflection. But although the collection of symmetries is not commutative, it is algebra, albeit noncommutative algebra. It satisfies many of the axioms for combining elements that characterize ordinary arithmetic but not all. Another accepted noncommutative system was the algebra of quaternions mentioned at the end of the last chapter. If it was relatively easy for people to accept noncommutative algebra, why was it so difficult for them to accept an alternate geometry, one that satisfied some but not all of the axioms of Euclid?

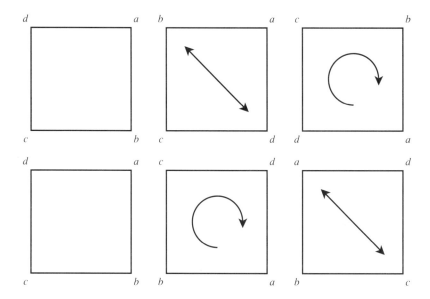

Reflecting across a diagonal then rotating a quarter turn (top) gives a different effect from rotating a quarter turn then reflecting (bottom).

One difference was that many mathematicians did not really believe that the axioms of geometry were independent. They recognized that the commutative property of algebra could not be proved from the other axioms, since in fact they had in front of them an example of a consistent system that satisfied all other properties except commutativity. But in geometry many thought that it was just a matter of time until someone was able to prove the validity of the fifth axiom from the first four axioms. It was quite a surprise when just the opposite happened— mathematicians formulated systems that looked like geometry in that they satisfied the first four axioms, but they failed to satisfy the fifth. No one had suspected that there could be a geometry having triangles whose angles summed to greater than or less than 180 degrees, and then these geometries appeared.

The Development of Non-Euclidean Geometry

The greatest mathematical thinker since the time of Newton was Karl Friedrich Gauss. In his lifetime, he revolutionized many different areas of mathematics, including number theory, algebra, and analysis, as well as geometry. Already as a young man, he had

Karl Friedrich Gauss, the greatest mathematical thinker of modern times, revolutionized the subject of geometry.

devised a construction for a 17-sided regular polygon using only the traditional Euclidean tools, the straightedge and compass. His most significant contributions to geometry came in his analysis of surfaces, and this analysis played an important role in the understanding of non-Euclidean geometrics.

Gauss was a surveyor, mapping out large areas of Europe, and he was an astronomer, studying phenomena in the heavens. In his land-based job, he triangulated areas, dividing them up into regions bounded by three of the shortest paths available on the surface of a sphere, namely great circle arcs. In his astronomical endeavors, he again used triangles to estimate distances, but this time the shortest paths available were the paths of light rays. In 1825 and again in 1827, Gauss combined insights from these fields to develop two ways of organizing information about surfaces, as intrinsic or extrinsic.

To appreciate the nature of the two approaches, he suggested a dimensional analogy—imagine what kind of geometry would be experienced by flatworms, two-dimensional creatures constrained to slide around on a surface. Although we three-dimensional beings are bound by gravity to spend most of our time moving along the surface of our planet, we can at least occasionally overcome that superficial limitation, as when we dig or jump to avoid obstacles. But the flatworm, unable to move up or down out of a surface, is virtually confined to a two-dimensional existence. How would an intelligent flatworm describe the geometry of its universe? If the surface were a flat plane like Flatland, the inhabitants would develop ordinary plane geometry, and as part of that geometry they would discover that the sum of the angles of any triangle was 180 degrees. But what if they lived on a very large sphere, so large that the curvature was not apparent to the creatures sliding along it? If they measured a small triangle, the angle sum would be close to 180 degrees, but for large triangles the results might be quite different. The geometry discovered by the flatworm would be the *intrinsic* geometry of a surface; this geometry depends only on those measurements made along the surface.

The *extrinsic* geometry of a surface depends on the way a surface sits in space, and that is the geometry we would see in looking down upon the flatworm's universe. Guided somewhat by his studies of astronomy, Gauss referred the geometry of the surface to the set of directions on the celestial sphere. Each point of a smooth surface has a closest approximating plane, the tangent plane. For each point on the surface, Gauss found a corresponding point on a unit sphere such that the tangent planes at the two points were parallel. By this means, Gauss defined what he called the spherical image mapping on a surface, one of the most powerful means of studying the way a surface curves in space.

The most surprising and powerful theorems of Gauss are the ones that relate the intrinsic and the extrinsic geometry of a surface. One of his biggest discoveries was that the extrinsic curvature related to the spherical image mapping could be determined from the intrinsic geometry, just by making measurements along the surface. The flatworm surveyor could discover crucial facts about the shape of its universe without ever leaving its superficial environment. Like ourselves, the flatworm would be able to assign a distance to any path and to define the distance between two points as the shortest length among all paths joining the points. Just as we measure angles between rays, it could measure the angle between a pair of shortest curves emanating from a point, and it could calculate the sum of the angles of a triangle just as we do. The flatworm's answer, however, might be totally different from ours. If we start with three points on the flatworm's universe, we can take a shortcut through space and connect them by straight line segments, and the angles of the triangle we obtain will sum to 180 degrees. The flatworm, on the other hand, might claim that the sum of the angles is not constant and that it depends on the size of the triangle.

Consider the particular case where the flatworm is confined to the surface of a sphere. He could construct a large triangle on the sphere running from the north pole to a point on the equator, then one-quarter of the way around the equator, and back to the north pole. Each of the angles of this spherical triangle is 90 degrees, so the angle sum is much greater than 180 degrees. For us who can leave the surface of the sphere to connect the points in space, the north pole and the two points on the equator form an equilateral triangle with each angle 60 degrees.

The geometric study of the sphere has a very long history, but by and large it was considered a subtopic in solid geometry. People spoke about great circle arcs, and they even knew in some sense that these represented the shortest paths on the surface of the globe. But they did not think of them as the same sorts of objects as the segments that provided the shortest distances in plane geometry. In ancient times, Ptolemy certainly knew that three great circle arcs forming a spherical triangle would determine angles adding up to more than 180 degrees, and in fact he

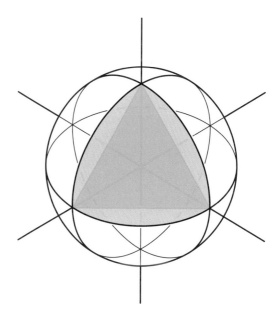

A spherical triangle can have three right angles between its pairs of great circle sides, even though the triangle with straight sides determined by the same vertices has angles of 60 degrees.

was able to prove that the bigger the area of the triangle, the larger the angle sum. With proper choice of units this relationship could be made explicit: the area of a triangular region on the sphere is precisely the amount by which its angle sum exceeds 180 degrees. Why didn't Ptolemy realize that this was an example of a non-Euclidean geometry, where the important Euclidean theorem that the angle sum equals 180 degrees simply does not hold? The answer is that he did not think of the relationships among points of a sphere and great circle arcs as a *geometry*. To qualify as a geometry, a system would have to have elements corresponding to points and lines, and the first four axioms would have to be satisfied. The system consisting of points on a sphere and lines given by great circle arcs did satisfy the third and fourth axiom, and even the second if we interpret it correctly, but it did not satisfy the first axiom. Although two nearby points on the sphere determine a unique great circle arc, there are point pairs for which this is not true. More than one great circle arc joins the north and south poles, and in fact there are infinitely many half-circles of longitude joining these two points, all of the same length. Thus spherical geometry did not qualify as a non-Euclidean geometry, although later on in this chapter we will see that it was closely related to one.

In the early part of the nineteenth century, mathematicians in three different parts of Europe found non-Euclidean geometries—Gauss himself, János Bolyai in Hungary, and Nicolai Ivanovich Lobachevsky in Russia. Each of them realized that it was possible to construct a two-dimensional geometry with points and shortest distance lines satisfying the first four axioms of Euclidean geometry, but not the fifth. The parallel postulate required that for any given point not on a given line, there is exactly one line through the point that does not meet the given line. There were two ways for this postulate to fail—if every line through the point meets the given line, or if there were two or more distinct lines through the point not meeting the line. The inventors of non-Euclidean geometry found systems based on both alternatives to the fifth axiom.

The alternative axiom stating that there could be more than one line through a given point not meeting a given line led to *hyperbolic geometry*. The theorems deduced by Bolyai and Lobachevsky seemed quite strange, but they were as consistent as Euclidean plane geometry. The diagrams that accompanied their demonstrations certainly did not look like those in Euclid's text, and mathematicians searched for some visual representation that

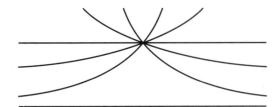

The alternative to the fifth axiom in hyperbolic geometry posits that through a point not on a given line, there are many lines not meeting the given line.

could make the new geometry easier to comprehend. One of the most successful expositors of this geometry, in England as well as in his native Germany, was the scientist Hermann von Helmholz. Appealing to the same thought experiment introduced by Gauss he used the dimensional analogy to explain a way of imagining a non-Euclidean two-dimensional geometry.

Helmholz asked his readers to consider a two-dimensional creature constrained to slide along the surface of a piece of marble statuary, measuring lengths of curves and sizes of angles. For example, a flatworm living on the surface of a cylindrical column would decide that for any region bounded by three shortest distance curves the angle sum would be 180 degrees, just the way it is on a plane, but if the column were in the shape of a long trumpet, the intrinsic geometry would be very different. The surface he suggested was a *pseudosphere*, invented by the Italian geometer Eugenio Beltrami. Although this surface had a sharp edge, it still illustrated most of the important properties of hyperbolic geometry, a geometry satisfying the first four axioms, but not the fifth. For any point and any shortest line, there were many lines through the point not meeting the line, and every triangle on the surface had an angle sum strictly less than 180 degrees!

The alternative axiom that every line through a given point would meet any other line led to *elliptic geometry*. This case was reminiscent of the geometry of the sphere, where every two great circles necessarily meet. The trouble with spherical geometry is that its straight lines meet twice. The radical solution was to throw away half the points of the sphere and just use the points in the southern hemisphere, below the equator. If the points of the southern hemisphere were the points of the geometry, and great circle arcs were the lines, then any two points did determine a unique line. The first axiom was saved. The third and fourth axioms still held, and the fifth certainly did not since there were triangles in the southern hemisphere having angle sums greater than 180 degrees.

But there was a new problem—the second axiom failed because the great semicircles lying in the southern hemisphere came to a dead stop when they reached the equator, and the second axiom requires that any line be indefinitely extendable. This difficulty was overcome by another radical suggestion: in addition to the points of the southern hemisphere, use half of the points of the equator, say those lying in the eastern hemisphere. When a point moving along a great circle arc in the southern hemisphere

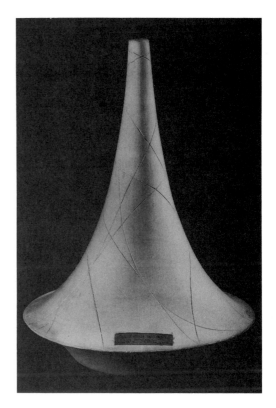

A nineteenth-century plaster model of Beltrami's pseudosphere.

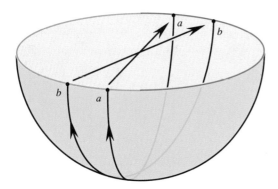

A model of elliptic geometry is given by the southern hemisphere with opposite points on the equator attached together.

Each point of the equator can be attached to its antipodal point by giving the equator a twist, then wrapping it around itself.

reaches the equator, it jumps instantly to the opposite point on the equator and continues to move along the great circle arc! In order for this procedure to work, we must consider the point on the equator at the prime meridian, with 0 degrees longitude, to be the same as the point on the international date line, with 180 degrees longitude. The amazing thing was that this idea worked. Using half the points on the sphere, it was possible to construct a geometry with great circle arcs as straight lines such that the sum of the angles of every triangle was greater than 180 degrees.

In elliptical geometry, it is as if every pair of antipodal points on the sphere represents the same point, and we only pay attention to the one lying in the southern hemisphere. The geometry is reminiscent of the geometry of lines through the origin in three-space, which we considered in Chapter 7. There we identified each line with the pair of antipodal points at which it meets the unit sphere. Thus the new interpretation of hemispherical geometry is associated with the geometry of lines, as well as with projective geometry.

This modification of spherical geometry had some weird consequences near the equator. In the new geometry, each point on the equator was considered to be the same as its antipode, directly on the other side of the sphere. When people tried to visualize this, they could easily attach opposite points together by wrapping the equator around itself, and they could even wrap together a small strip made by including part of the surface near the equator. But in order to do this it was necessary to introduce a twist, and when people tried to extend the construction over the entire southern hemisphere, they failed. The new geometry worked all right—it did not depend on being able to put it together in three-dimensional space. Fortunately by this time mathematicians knew of a place where the new geometry could be built, and that was in the fourth dimension.

Three-Dimensional Non-Euclidean Geometry

Bolyai, Lobachevsky, and Gauss had created two-dimensional non-Euclidean geometries. For any point, the surrounding space looked like a piece of the plane. To check on the possible curvature of the space it might suffice to make some very careful measurements. In fact if the curvature of the space is too gradual or if we draw too small a triangle, we might not realize that the intrinsic geometry is non-Euclidean. To see this, we only have to go back to Gauss and his activity as a surveyor. If he confines himself to small enough regions, then the angle sum will indeed be 180 degrees, within the tolerance of his measuring instruments. But for a large enough triangle the difference is appreciable.

The idea that there could be different kinds of geometries on surfaces was strange enough, but even more threatening to convention was the suggestion that there could be different kinds of three-dimensional geometry. Surely there was only one way to think of space? At least that is the firm opinion stated by the followers of Immanuel Kant, who felt that any alternative was unthinkable. But it wasn't unthinkable for Gauss. He not only considered the possibility of non-Euclidean three-space, he also speculated about whether or not this non-Euclidean model might be the true description of the space we live in.

Is it possible that our space is curved rather than flat? One of Gauss' most important insights was that we can tell the shape of the space we are in by measuring angle sums for triangles, not just in two dimensions, but also in three-dimensional space. To show that space is non-Euclidean, all we have to do is find a triangle with an angle sum observably different from 180 degrees. Gauss set out to measure the angle sum for the largest triangle he could find. He did not want to lay it out along the surface of the earth, where he knew that angle sums of spherical triangles could be greater than 180 degrees. Instead he used what he thought of as the straightest lines in space, represented by light rays. In order to make a large triangle of light rays, he positioned beacons on the tops of three high mountains, where the curvature of the earth would not block the light rays from the sight of observers positioned on each mountain. The observers measured the angles and totaled up the answers, but the experiment was inconclusive. The sum was 180 degrees, up to the accuracy of their surveying instruments. It is quite difficult to prove that the sum of the angles of a triangle of light rays is precisely equal to 180 degrees. Even a modern computer cannot establish that two numbers are exactly equal, although it can check easily enough that two numbers are equal up to a desired tolerance.

We still do not know whether or not our three-dimensional space satisfies the axioms of Euclid's solid geometry, but we do know that we can't use light rays as our straight lines in this geometry. One of the crucial discoveries of physics is that light rays are deflected as they pass near a very massive object. Thus a ray might bend as it passed a star, and the bend would alter the angle sum of a triangle of light rays. That does not mean that our geometry is a non-Euclidean three-dimensional geometry, but it does mean that we have to be careful in trying to apply such a geometry to the study of light rays traveling interstellar distances.

Higher-Dimensional Euclidean Geometries

The ideas of non-Euclidean geometry became current at about the same time that people realized there could be geometries of higher dimensions. Some observers lumped these two notions together and assumed that any geometry of dimension higher than three had to be non-Euclidean. But soon mathematicians realized that there was an essential difference between the two notions. It was perfectly possible to have a higher-dimensional geometry satisfying axioms that were exactly analogous to all the axioms of Euclid, so that any triangle would have its angle sum precisely equal to 180 degrees.

Hermann Grassmann in Germany was one of the first to develop a full geometry that worked in dimensions higher than three, and this notion was extended in England by Arthur Cayley and John J. Sylvester among others. In particular they described a geometry of four dimensions where the fundamental objects were points, lines determined by pairs of points, planes determined by noncollinear triples of points, and hyperplanes determined by noncoplanar quadruples of points. They were able to move into a higher dimension because they added another axiom: outside any given three-dimensional hyperplane, there were other points.

In solid geometry, the analogue of the parallel axiom says that through any point not on a plane, there passes exactly one plane that does not meet the first plane. In four-dimensional geometry, the analogous axiom states that through any point not lying on a hyperplane, there passes exactly one hyperplane not meeting the first hyperplane.

In plane geometry, through a given point on a line there is a unique perpendicular line. In solid geometry, through a given point on a line there are many lines that meet it at a right angle, and these lines all lie in a plane perpendicular to the given line through the point. In four-dimensional geometry, the lines that meet a given line at a given point at a right angle fill out an entire hyperplane.

In solid geometry, through a point on a plane there passes exactly one line perpendicular to the plane. In four-dimensional geometry, through a point on a plane there are many lines that meet the plane at a right angle, and these lines fit together to form a plane perpendicular to the first plane and meeting the plane at a single point.

It is quite surprising that in four-space two totally perpendicular planes can meet at a point (see diagram on page 144). This concept is so difficult precisely because it is so hard to visualize. But in fact it is not that easy to see that a line in three-space is perpendicular to a plane. If the plane is opaque, then the line appears to come down to the plane and disappear beneath it. If the plane is semitransparent, we might notice that the line changes color as it goes through. We could move the configuration around to determine if the line is really perpendicular. If we draw a square on the plane and hold the plane so that the square appears as a square, with four equal sides and four right angles, then the perpendicular line will appear as a point. We can do the same in four dimensions with the help of a graphics computer. We can move a configuration of two perpendicular planes around in four-space so that the image on the screen is just a plane, with the other plane being mapped to a point. A good place to find a treatment of results in synthetic geometry of four dimensions is in the book *Geometry of Four Dimensions* by Henry Parker Manning.

Eventually the fundamental ideas of Gauss and the development of higher-dimensional analytic geometry led to a beautiful general theory in the dissertation of Bernhard Riemann, who conceived the notion of an *n*-dimensional manifold with a metric, a rule for assigning lengths to paths. The far-reaching effects of this generalization have changed the way mathematicians view the nature of space, and they formed the necessary background for the appearance of relativity physics. Higher-dimensional analytic

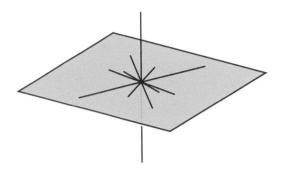

There are many perpendicular lines through a point on a line in space, and these lines fill out the plane perpendicular to the given line through the point.

Henry Parker Manning of Brown University wrote *Geometry of Four Dimensions* in 1914 and edited *The Fourth Dimension Simply Explained* in 1910.

methods led some mathematicians to adopt a purely formal approach to geometry, independent of the traditional ways of visualizing geometric objects. But from another point of view, these methods laid the groundwork for the use of computer graphics in investigating objects in higher-dimensional space.

Immanuel Kant and Nonorientability

It was not only the emergence of new axiom systems of non-Euclidean geometry that upset the followers of Immanuel Kant. Another serious point of contention centered around a different geometric notion, *orientability*. In the plane we say that two figures can be superposed if it is possible to slide one to the exact position occupied by the other. Imagine two figures drawn on transparent plastic sheets able to glide over one another. The two figures could be superposed if we could position one directly on top of the other.

According to traditional geometry, if two figures could be superposed, they were directly congruent, but not every pair of congruent figures could be superposed merely by sliding. One of Euclid's main theorems about triangles says that if the lengths of the sides of one triangle equal those of another, then the two triangles are congruent. But consider a right triangle with three different side lengths and its reflected image in a line (the *Flatland*

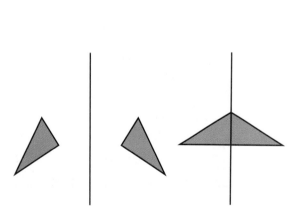

When a scalene triangle is reflected across a line (left), the triangles cannot be superposed in the plane (right).

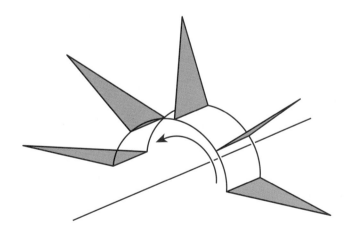

By rotating a scalene triangle about a line in space, we can superpose it on its reflected image.

equivalent of a full-length mirror). Corresponding sides of the object have the same edge lengths, but it is impossible to superpose them by sliding along the plane. Indeed, if we slide one of the triangles so that two edges of the same length come together, then the two triangles lie on opposite sides of the line containing their common edge.

Kant would say that these two triangles are an *enantiomorphic pair*, congruent but not superposable. It is clear that this notion depends in an essential way on the geometry of the plane. If we consider the two triangles as situated in space, then it is quite easy to pick one up and place it down on the other, like turning the pages of a book.

It follows that the definition of enantiomorphic pair depends on the dimension of the space we are working in. This fact becomes crucial when we consider enantiomorphic pairs of objects in three-dimensional space. Instead of a right triangle, consider an off-center pyramid cut from a corner of a rectangular parallelepiped having edges of three different lengths. We can cut off another corner to obtain an off-center pyramid that has the same side lengths for each of its edges. The corresponding triangles in the two pyramids are congruent, but still the pyramids themselves are not superposable. They are mirror images of one another, and they have different "handedness."

The example Kant used was a marble hand broken off a statue. We can easily determine if the hand is a right or a left by attempting to shake hands with it with our right hand. If we succeed, the marble hand is also a right hand; otherwise it is a left hand. But what if we can't get close to the hand? How can we determine whether it is a right hand or a left? Kant set up a more extreme thought experiment—what if the marble hand were the only object in the universe? Would it make sense to say that it is still either a right hand or a left?

In a sense the problem vanishes completely if we do not restrict ourselves to three-dimensional space. In the same way that a triangle can be turned into its mirror image by a rotation about a line in three-space, a marble hand can be turned into its mirror image by a rotation about a plane in four-space. Just as it does not make sense to ask whether a free-floating cutout of a handprint, on a sheet of paper with both sides identical, is a right or a left handprint, so we cannot make sense of the question of whether a marble hand floating freely in four-space is right or left. This answer does not please the followers of Kant even to this day.

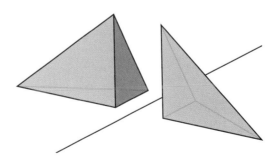

Mirror-image off-center pyramids are congruent but not superposable.

Möbius Bands, Real Projective Planes, and Klein Bottles

There is another way to create nonorientable objects, not by changing the dimension but by altering the shape of the space. In 1840 August Möbius invented the surface bearing his name, the *Möbius band*. It is constructed by pasting together the two vertical edges of a long rectangle, but with a twist so that the top vertex on one side is connected to the bottom vertex on the other. The resulting surface has only one edge. We should think of the band as made of some porous material through which an ink-drawn figure bleeds, leaving us no way to tell on which side of the strip the figure was originally placed. A triangle drawn on, or rather *in*, this two-dimensional strip can slide along the strip and come back so that it is superposed on its mirror image. There are no enantiomorphic pairs on a Möbius band.

The Möbius band is an example of a *nonorientable* space, which means that it is not possible to distinguish an object on the

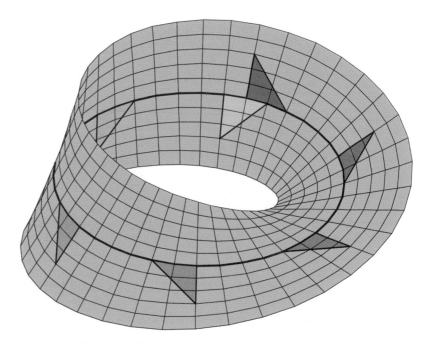

On a Möbius band, a scalene triangle and its reflected image can be superposed.

surface from its reflected image in a mirror. A surface will be nonorientable precisely when it contains one of these orientation-reversing paths. It would be quite a shock to the inhabitants of Flatland if some explorer returned from an expedition with all his right-handed tools transformed into left-handed ones. This scenario is developed at length in an excellent treatment of alternate geometries called *The Shape of Space*, by Jeff Weeks, giving an introduction to the research program of William Thurston. Since we do not really know all that much about the large-scale structure of our universe, it might be possible for some future interstellar explorer to discover an orientation-reversing path in our own three-dimensional space, permitting all of our monkey wrenches and marble right hands to be transformed into left-handed objects. This would dismay the future followers of Kant.

Two important surfaces containing Möbius bands can be built in four-dimensional space. Both have the important property of being without a boundary, like the surface of a sphere. The first, called a *real projective plane*, is obtained by attaching the boundary of a disc to the boundary of a Möbius band. The second is formed by attaching two Möbius bands along their common boundary to form a nonorientable surface called a *Klein bottle*, named for its discoverer, Felix Klein.

We have encountered the real projective plane earlier in this chapter in our discussion of elliptic geometry. There we started with the southern hemisphere and we had to imagine attaching together opposite points on the equator. As we can see by considering a thin strip around a great circle arc through the south pole, the ends of the strip on the equator must be attached with a twist, forming a Möbius band. The remainder of the southern hemisphere consists of two half-discs that fit together to form a disc with its boundary attached to the boundary of the Möbius band. Thus, the space of elliptic geometry—the southern hemisphere with each point on the equator attached to its opposite point—can be described as a disc attached to a Möbius band, and therefore as a real projective plane.

One of the easiest ways to imagine building any surface is to picture it made out of triangles. Instead of using a hemisphere of a round sphere, we can attempt to build a real projective plane by using the ten triangles in half a regular icosahedron. This half icosahedron has a boundary consisting of six edges, and each edge is to be attached to the opposite edge with a twist, leaving a boundary of three vertices. As there are also three visible vertices not on the boundary, we obtain a representation of the real projective plane using just six vertices and ten triangles. A strip of five triangles stretching from one boundary segment to the opposite segment forms a Möbius band when its end segments are attached with a twist. The Möbius band contains five of the six vertices, and the remaining five triangles fit together to define a disc around the sixth vertex.

In order to try to construct the real projective plane in three-space, we can build a five-vertex Möbius band by choosing five triangles in the projection of a four-simplex, as indicated in the illustration below. The boundary of this band will consist of a pentagon in space. In order to attach the remaining five triangles without running into the triangles already placed, we have to locate a point from which we could see all five edges of the pentagon, unobstructed by previously chosen sides. It turns out to be impossible to find such a viewing point in three-dimensional space, but it is easy if we are willing to go to four-space. Just as we in three-space can see all rooms of a Flatland house, from a four-dimensional vantage point we could see all points in three-space simultaneously. Thus we can form the five remaining triangles without running into any triangles of the band by connecting all five edges of the boundary of the Möbius band to a point in four-space. Therefore it is possible to construct a real projective plane

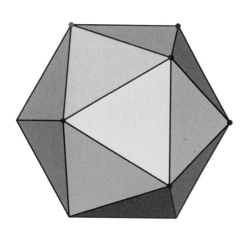

The real projective plane can be thought of as half an icosahedron with opposite segments attached with a twist.

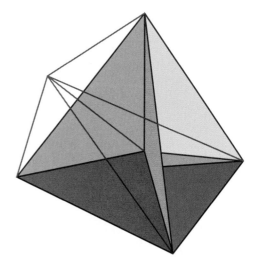

To construct the real projective plane, connect the boundary of a five-triangle Möbius band to a point.

in four-space, although it cannot be done in three-dimensional space. We can obtain an even more symmetrical example of a real projective plane by taking ten of the equilateral triangles determined by the six vertices of a regular five-simplex embedded in five-dimensional space. A powerful theorem states that no surface without a boundary and containing a Möbius band can be constructed without self-intersection in three-space. The reason is rather subtle, but it becomes clearer if we use a dimensional analogy. In a Möbius band, it is possible to find two closed curves that cross at one point, something that is impossible in the plane where pairs of curves always cross at an even number of points. Therefore we can conclude that it is impossible to construct a Möbius band in the plane. Now for a Möbius band in three-space, it is possible to follow along very close to the center curve and end up on the other side of the band, just opposite the starting point. If we then pierce through the band with a segment, we obtain a closed curve in space that meets the band at one point. If the band were part of a surface without boundary, like a Klein bottle or a projective plane, then this construction would yield a curve in three-space intersecting such a surface in one point. But in three-space any closed curve must intersect any surface without boundary in an even number of points. Therefore it is impossible to construct any boundaryless nonorientable surface without self-intersection in ordinary three-space.

Another extremely important nonorientable surface is the Klein bottle. The instructions for building a Klein bottle are simple enough: start with a rectangle and attach the vertical sides with a twist and the horizontal sides without a twist. We can follow one instruction or the other but not both. Since the instructions describe a surface without a boundary but containing a Möbius band, we know by the argument in the previous paragraph that we cannot construct a Klein bottle without self-intersection in three-space. The reader will not be surprised to learn, however, that we can construct one in four-space. One way is to start with the five-vertex Möbius band described above and to move it in a direction perpendicular to three-space until it reaches a parallel Möbius band. During this movement, the five edges in the pentagonal boundary of the Möbius band trace out five squares joining the boundary of one Möbius band to the other, thus forming a Klein bottle constructed in four-space without self-intersection.

For many years, people attempted to build representations of Klein bottles without self-intersection in ordinary space, and in

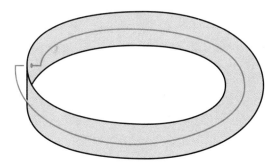

A closed curve in space near the center of a Möbius band meets the band an odd number of times.

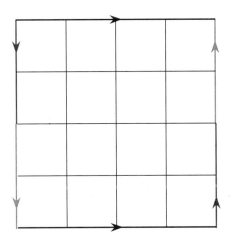

The instructions for a Klein bottle are to attach the top and bottom edges of a rectangle to form a cylinder, and to attach the left to the right side with a twist to form a Möbius band.

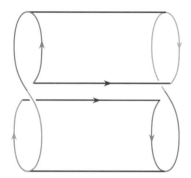

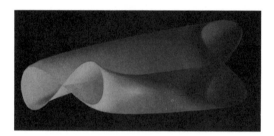

Top: Folding the top edge of the rectangle forward and the bottom edge backward forms a figure eight cylinder, passing through itself along a segment. *Bottom:* To produce a figure eight Klein bottle from the figure eight cylinder, we would twist the figure eights as we bring the two ends outward. This computer-generated image shows the Klein bottle before the final attachment has been made.

each case they had to make a compromise and allow the surface to pass through itself. Glassblowers fashioned Klein bottles in space by allowing a tubular portion of the surface to intersect a ring-shaped area on the surface. The sculpture that opens this chapter is a fine recent example of this construction. Another construction was suggested by our work in four-dimensional geometry, where we start to build the Klein bottle not by making a round cylinder but by allowing the cylinder to pass through itself to form a "figure eight" cylinder. We can then attach the figure eights at the ends with a twist to accomplish the identification required to make a Klein bottle. When modeling surfaces on a computer screen, it turns out to be easier to describe this figure eight Klein bottle than the glassblower's version. The modern graphics computer makes it possible for us to design these intriguing objects in four-dimensional space and then investigate them by projecting them into our own viewing space, a fitting final insight in our project of visualizing dimensions.

A computer graphics representation of the figure eight Klein bottle, projected from four-space.

FURTHER READINGS

An excellent reference for the relationship of physics and dimensions is the recent Scientific American Library volume, *A Journey into Gravity and Spacetime* by John Archibald Wheeler (W. H. Freeman, 1990). Two books by Rudolf Rucker also present the relationship between geometry and relativity: *The Fourth Dimension: Toward a Geometry of Higher Reality* (Houghton-Mifflin, 1984), and *Geometry, Relativity, and the Fourth Dimension* (Dover, 1977). The former book also treats the philosophical and mystical side of the subject, not only through Rucker's own ideas, but by incorporating the writings of Charles Howard Hinton, whose work Rucker collected in *Speculations on the Fourth Dimension: Selected Writings of C. H. Hinton* (Dover, 1980). Jeff Weeks has written a fine account of current work in the topology of three-dimensional spaces in *The Shape of Space* (Marcel Dekker, 1985).

The ideal resource for the history of geometry in the nineteenth century is the recent book by Joan Richards, *Mathematical Visions: The Pursuit of Geometry in Victorian England* (Academic Press, 1988), which contains an excellent bibliography. A historical survey of higher dimensions appears in the introduction to *Geometry of Four Dimensions*, written by Henry Parker Manning in 1914 and reprinted by Dover in 1956. See also the preface of Manning's book *The Fourth Dimension Simply Explained* (Munn and Company, 1910; reprinted by Dover, 1960).

The definitive book on higher dimensions and art is Linda Henderson's volume *The Fourth Dimension and Non-Euclidean Geometry in Modern Art* (Princeton University Press, 1983). Another good resource on the subject is *Hypergraphics: Visualizing Complex Relationships in Art, Science, and Technology*, edited by David Brisson and containing the pioneering paper of A. Michael Noll as well as many other articles (Westview Press, 1978).

Martin Gardner has featured the fourth dimension in many of his *Scientific American* columns, and these have been anthologized in several volumes of his collected works, including *The Unexpected Hanging* (Simon and Schuster, 1986), *Mathematical Carnival* (Alfred A. Knopf, 1975), and *The New Ambidextrous Universe*

(W. H. Freeman, 1990). Alexander Dewdney has also written articles on dimensions in his *Scientific American* columns. Particularly interesting is his treatment of two-dimensional science and technology, which is discussed at length in his allegory *The Planiverse* (Poseidon Press, 1984). Dewdney's column "Computer Recreations" in the April, 1986, issue of *Scientific American* featured the animation of the hypercube.

Ivars Peterson has included in his book *The Mathematical Tourist* (W. H. Freeman, 1988) a report of the conference on hypergraphics at Brown University on the occasion of the centennial of *Flatland* in 1984. There have been several new editions of *Flatland* during the past ten years, and the best-selling Dover edition, first printed in 1952, has a good introduction by Banesh Hoffman. A new edition by Princeton University Press appeared in Spring, 1991, with an introduction written by Thomas Banchoff. A sequel to *Flatland* called *Sphereland*, by Dionys Burger, was first published in 1965 and was reprinted by Harper and Row in 1983.

For fractal geometry there are several very good sources, including *The Fractal Geometry of Nature* by Benoit Mandelbrot (W. H. Freeman, 1982), *The Beauty of Fractals* by Heinz-Otto Peitgen and P. H. Richter (Springer-Verlag, 1986), and *Fractals Everywhere* by James Barnsley (Academic Press, 1988).

General references on the topology of surfaces include the classic *Geometry and the Imagination* by David Hilbert and Stefan Cohn-Vossen (Chelsea, 1952) and *A Topological Picturebook* by George Francis (Springer-Verlag, 1988).

The specific reference for the dynamical systems related to pendulum motion is the paper "Topology and Mechanics" by Hüseyin Koçak, Fred Bisshopp, David Laidlaw, and the author in *Advances in Applied Mathematics*, Vol. 7, pages 282–308, 1986. A good reference for singularity theory in geometry is *Curves and Singularities* by Peter Giblin and James Bruce (Cambridge University Press, 1984).

For polyhedra the standard references are the books of H. S. M. Coxeter: *Regular Polytopes* (Dover, 1973), *Regular Complex Polytopes* (Cambridge University Press, 1974), and *Introduction to Geometry* (John Wiley and Sons, 1961). The classic book by D. M. Y. Sommerville, *Geometry of n Dimensions* (Methuen, London, 1929; reprinted by Dover, 1958) is also an excellent reference. Another, more formal source is *Convex Polytopes* by Branko Grünbaum (Interscience Publishers, 1967). More recent books are Arthur Loeb's *Space Structures: Their Harmony and Counterpoint*

(Addison-Wesley, 1976) and *Shaping Space*, edited by Marjorie Senechal and George Fleck (Birkhäuser, 1988). Another reference for the coordinate geometry of four dimensions is *Linear Algebra through Geometry* by Thomas Banchoff and John Wermer (Springer-Verlag, 1983, second edition, 1991).

The premier works on the visualization of data are Edward Tufte's *The Visual Display of Quantitative Information* (Graphics Press, 1983) and John Tukey's *Exploratory Data Analysis* (Addison-Wesley, 1977). Also by John Tukey and cowritten by Paul Tukey is a fine article "Graphic Display of Data Sets in III or More Dimensions," which appeared in *Interpreting Multivariate Data* (John Wiley and Sons, 1981).

Several works of fiction are extremely insightful, including Madeleine L'Engle's *A Wrinkle in Time* (Farrar, Straus, and Giroux, 1962), which introduced generations of readers to the concept of the tesseract. Robert Heinlein's story ". . . and He Built a Crooked House" appears in Clifton Fadiman's collection *Fantasia Mathematica* (Simon and Schuster, 1958), along with other mathematical stories.

Finally, *The Hypercube: Projections and Slicing* is available on film and videotape from International Film Bureau, 332 South Michigan Avenue, Chicago, Illinois 60604. Videotapes of *The Hypersphere: Foliation and Projections* and *Fronts and Centers* are distributed by The Great Media Company, P. O. Box 598, Nicasio, CA 94946. For ongoing information about the author's visualization projects, contact http://www.geom.umn.edu/~banchoff on the Internet. Professor Banchoff's home page contains many examples of interactive computer graphics illustrations related to the material in *Beyond the Third Dimension*.

ACKNOWLEDGMENTS

I would like to thank a number of people who have provided me with inspiration, assistance, and support over the years and who have contributed directly or indirectly to the writing of this book.

My earliest mathematical encouragement came from my mother, a kindergarten teacher, and my father, a payroll account-ant. Herb Lavine, three years older than I, taught me algebraic patterns on the tops of cartons in his father's grocery store. In my freshman year at Trenton Catholic Boys' High School, Father Ronald Schultz listened to my first theorem—calculating when advancing shadows would bisect the triangular tiles on the church floor—and he also paid attention to my first theory relating the fourth dimension and the Trinity. William Hausdoerffer at Trenton State gave the first college lecture I ever heard, and today he remains my consultant on the subject of sundials.

At the University of Notre Dame my advisors, and eventually my colleagues and friends, were Frank O'Malley in writing, R. Catesby Taliaferro in mathematics, Dean Charles Sheedy, Arnold Ross, who gave me my first chance to teach at his summer program for high school teachers, G. Y. Rainich, who introduced me to non-Euclidean geometry and relativity, and Father Robert Pelton and Father John Dunne, who gave me the opportunity to explore geometry and theology.

My Ph.D. advisor at the University of California, Berkeley, Professor Shiing-Shen Chern, and my primary geometric advisor, Nicolaas Kuiper, encouraged me while I studied the geometry of polyhedral surfaces in higher dimensions. As a Benjamin Peirce Instructor at Harvard I had my first chance to teach the fourth dimension. I appreciate my contact there with William Reimann and William Wainwright at the Carpenter Center for the Visual Arts, and with my friends Edwin Moise and George Hunston Williams.

At Brown University I thank particularly Charles Strauss, my collaborator in computer graphics and geometry for more than twelve years, and Harold Weber, who built the first device that let us see four-dimensional objects in real time. Many Brown University colleagues have contributed directly to the development of the ideas in this book, including Fred Bisshopp and Philip Davis (applied mathematics), Hunter Dupree (history), Richard Fishman (art), Richard Gould (anthropology), Peter Heywood (biology), John Hughes (mathematics), Hüseyin Koçak (applied mathematics), Martha Mitchell (archives), Doctor Alfred Moon (radiology), Henry Pohlmann (mathematics), Joan Richards (history), Karen Romer (dean), James Schevill (English), Gerald Shapiro (music), Merton Stoltz (provost), Julie Strandberg (dance), James Van Cleve (philosophy), Andries van Dam (computer science), Tom Webb (geology), and Arnold Weinstein (comparative literature). I am grateful for the constant help of the mathematics office staff—Dale Cavanaugh, Carol Oliveira, and Natalie Johnson (and her father). Thanks also to my colleagues at different institutions—Antony Raubitschek (classics), Joan and Erik Erikson (psychology), and John Tukey and Paul Tukey (exploratory data analysis), and the members of the Clavius Group. I would like to acknowledge the National Science Foundation, the Office of Naval Research, and the Mathematical Sciences Education Board for their support of several projects described in the book. My geometric collaborators Peter Giblin, Wolfgang Kühnel, Ockle Johnson, and Clint McCrory were patient while I worked on this book. I remember especially my late geometry colleagues Hassler Whitney, William Pohl, and Stephanie Troyer.

All artists involved with dimensions in art acknowledge the contributions of David Brisson at the Rhode Island School of Design and founder of the Hypergraphics Group. I thank him and the other artists whose work appears in this book: James Billmyer, Salvador Dalí, Attilio Pierelli, Lana Posner, Tony Robbin, and José Yturralde. I would also like to thank artists Nieves Billmyer, Harriet Brisson, Arthur Loeb, Colin Low, Michele Emmer, Charles Eames, Malcolm Grear, C. C. Beck (who drew the Captain Marvel comic that started it all), and all my friends at the Providence Art Club, especially David Aldrich, William Gardner, Carlton Goff, Garvin Morris, Maxwell Mays, Raymond Parker, and Thomas Sgouros. I acknowledge the influence of many conversations with my fellow writers about higher dimensions, including Dionys Burger, H. S. M. Coxeter, Alexander Dewdney, Henry Thomas Dolan, Martin Gardner, Linda Dalrymple Henderson, Madeleine L'Engle, and Jeff Weeks. Special help for this project came from my friends in England—Sir Basil Blackwell, William Hallett and Terry Heard at the City of London School, Barbara Phillipson, and David and Deborah Singmaster. Many friends have listened to me talk about this book over the years, in particular Donald Albers, Doctor Frederick Barnes, Carl Bridenbaugh, Daniel Driscoll, James Fitzwater, Ambrose Kelly, Margaret Langdon Kelly, Judge William Mackenzie, Louise Mackenzie, David Masunaga, Thomas Roberts, and Allen Russell, and during this past year, Peter Chase, Harold Ellsworth, Bishop John Higgins, and Robert Morehead Perry.

In the illustration acknowledgments I will individually credit the efforts of the many students who have contributed to our work in geometry and computer graphics at Brown University. Finally I would like to thank all the students who have participated in my courses on the fourth dimension over the past twenty-five years, and especially those who have been my assistants for those courses, including Michael Holleran, Steven McInnis, Lindley Gifford, David Pinchbeck, Brandt Goldstein, David Goldsmith, Anne Morgan, Michael Chorost, Ilise Lombardo, Eric Chaikin, and now David Burrowes and Matthew Salbenblatt. They will recognize many of their own ideas in this book.

ILLUSTRATION ACKNOWLEDGMENTS

A number of individuals and groups must be acknowledged for creating the many illustrations in this book.

Nicholas Thompson, a Brown University undergraduate, and the author produced the computer-generated images on pages 4, 123, 126 to 129, 147, 148 (top), 149, 150, 151 (bottom), 152 to 154, 174, and 175 during a two-year project at Prime Computer, Inc., in association with Robert Batchelder, Greg Berghorn, and Robert Gordon. The program was run on a Prime PXCL 5500 workstation. The computer-generated Klein bottles on pages 11 and 198 were also rendered by Nicholas Thompson on the Prime workstation, following a design developed by David Salesin.

Applied Mathematics professors Hüseyin Koçak and Fred Bisshopp, computer science graduate students David Laidlaw and David Margolis, and the author worked on a joint project that created the computer-generated images on pages 125, 136, 137, 169, and 177. The images were produced on a Lexidata Solidview computer, using software developed at Brown University.

The computer-generated images on pages 145, 148 (bottom), 151 (top), and 188 were produced using the computer program Vector, an interactive program developed by Rashid Ahmad and extended by Jeff Achter, Cassidy Curtis, Curtis Hendricksen, Greg Siegle, and Matthew Stone. The program runs on a SUN workstation.

The images on page 33 were generated using an implicit function program. The program was developed by Kevin Pickhardt and Steven Ritter, modified by Trey Matteson, and run by Edward Chang.

Other students whose early work served as a basis for the more recent programs include Andrew Astor, Timothy Kay, Edward Grove, Richard Hawkes, Robert Shapire, Kathleen Curry, Richard Schwartz, Paul Strauss, Steve Feiner, and Scott Draves.

The line drawings in Chapters 3 and 4 were rendered by Nicholas Thompson. While a mathematics graduate student at Brown University, Davide Cervone created nearly all of the line drawings using Aldus Freehand on a Macintosh II computer. For this paperback edition, he has updated the images on pages 56 to 61 utilizing new rendering techniques at The Geometry Center, University of Minnesota, Twin Cities campus.

SOURCES OF ILLUSTRATIONS

facing page 1
George Wright, Widenfeld and Nicolson, Ltd.

page 11
top, Deutsches Museum

page 12
Robert Ivy/Ric Ergenbright Photography

page 14
Keith H. Murakami/Tom Stack and Associates

page 21
top, Byron Crader/Ric Ergenbright Photography

page 31
From *The Rhind Mathematical Papyrus*, edited by A. B. Chace and H. P. Manning, Mathematical Association of America, 1927 and 1929.

page 36
Ric Ergenbright Photography

page 38
Dr. Alfred Moon, Rhode Island Hospital

page 39
top, City of London School Archives

page 41
Dr. Alfred Moon, Rhode Island Hospital

page 46
top, Susan Schwartzenberg/ Exploratorium

page 57
bottom, Louise Morse

page 62
Scala/Art Resource

page 66
left, Donald P. Greenberg
right, generated on an Evans and Sutherland frame buffer by Peter Atherton, Kevin Weiler, and Donald P. Greenberg at the Cornell Program of Computer Graphics, 1977.

page 69
Tony Robbin

page 78
top, Joan W. Nowicke, Smithsonian Institute
bottom, from T. Webb III, E. J. Cushing, and H. E. Wright, Jr., "Holocene Changes in the Vegetation of the Midwest." *Late-Quaternary Environments of the United States*, edited by H. E. Wright, Jr., Vol. 2, page 151, University of Minnesota Press, Minneapolis, 1983.

page 79
From T. Webb III, E. J. Cushing, and H. E. Wright, Jr., "Holocene Changes in the Vegetation of the Midwest." *Late-Quaternary Environments of the United States*, edited by H. E. Wright, Jr., Vol. 2, page 154, University of Minnesota Press, Minneapolis, 1983.

page 81
From T. Webb III, E. J. Cushing, and H. E. Wright, Jr., "Holocene Changes in the Vegetation of the Midwest." *Late-Quaternary Environments of the United States*, edited by H. E. Wright, Jr., Vol. 2, page 153, University of Minnesota Press, Minneapolis, 1983.

page 82
Tom Webb and Sarah Stead

page 83
Tom Webb and Sarah Stead

page 84
Ric Ergenbright Photography

page 100
From H. S. M. Coxeter, *Regular Complex Polytopes*, Cambridge University Press, 1974.

page 101
top, The Franklin Institute Science Museum, Philadelphia

page 104
right, José Yturralde

page 105
Metropolitan Museum of Art, gift of the Chester Dale Collection, 1955 (55.5).

page 108
Chaz Nichols

page 110
courtesy of the author

page 111
Art Resource

page 112
top, Lana Posner

page 113
Neil Leifer/Sports Illustrated

page 115
bottom, Attilio Pierelli

page 118
from David Hilbert and Stefan Cohn-Vossen *Geometry and the Imagination*, Chelsea, 1952.

page 119
David Brisson, Rhode Island School of Design

page 120
Stanford University Museum of Art, 41.1018, Muybridge Collection

page 130
The Vermont Rehabilitation Engineering Center for Low Back Pain

page 133
The Vermont Rehabilitation Engineering Center for Low Back Pain

page 134
courtesy of the author

page 135
Carl Yarborough

page 138
Richard Gould

page 139
Adapted from Richard A. Gould and
John E. Yellen, "Man the Hunted."
Journal of Anthropological Archaeology,
Vol. 6, pages 17–103, 1987.

page 142
© The Phillips Collection, Washington,
D.C.

page 156
courtesy of the artist

page 160
top, Richard Schoenbrun

page 169
top, Dwight Kuhn

page 178
Klein bottle by Dr. William D. Clark III,
El Segundo, CA; photo by Chip Clark

page 180
top, John Hay Library, Brown University

page 181
Deutsches Museum

page 184
Deutsches Museum

page 187
From Gerd Fischer, *Mathematical Models,*
Vieweg-Verlag, 1986.

page 191
bottom, Brown University Archives

INDEX

Selected hardcover books in the Scientific American Library series:

ATOMS, ELECTRONS, AND CHANGE
by P. W. Atkins

DIVERSITY AND THE TROPICAL RAINFOREST
by John Terborgh

GENES AND THE BIOLOGY OF CANCER
by Harold Varmus and Robert A. Weinberg

MOLECULES AND MENTAL ILLNESS
by Samuel H. Barondes

EXPLORING PLANETARY WORLDS
by David Morrison

EARTHQUAKES AND GEOLOGICAL DISCOVERY
by Bruce A. Bolt

THE ORIGIN OF MODERN HUMANS
by Roger Lewin

THE EVOLVING COAST
by Richard A. Davis, Jr.

THE LIFE PROCESSES OF PLANTS
by Arthur W. Galston

IMAGES OF MIND
by Michael I. Posner and Marcus E. Raichle

THE ANIMAL MIND
by James L. Gould and Carol Grant Gould

MATHEMATICS: THE SCIENCE OF PATTERNS
by Keith Devlin

A SHORT HISTORY OF THE UNIVERSE
by Joseph Silk

THE EMERGENCE OF AGRICULTURE
by Bruce D. Smith

ATMOSPHERE, CLIMATE, AND CHANGE
by Thomas E. Graedel and Paul J. Crutzen

AGING: A NATURAL HISTORY
by Robert E. Ricklefs and Caleb E. Finch

INVESTIGATING DISEASE PATTERNS: THE
SCIENCE OF EPIDEMIOLOGY
by Paul D. Stolley and Tamar Lasky

Other Scientific American Library books now available in paperback:

POWERS OF TEN
by Philip and Phylis Morrison and the Office of Charles and Ray Eames

THE DISCOVERY OF SUBATOMIC PARTICLES
by Steven Weinberg

THE SCIENCE OF MUSICAL SOUND
by John R. Pierce

THE SECOND LAW
by P. W. Atkins

MOLECULES
by P. W. Atkins

THE NEW ARCHAEOLOGY AND THE ANCIENT MAYA
by Jeremy A. Sabloff

THE HONEY BEE
by James L. Gould and Carol Grant Gould

EYE, BRAIN, AND VISION
by David H. Hubel

PERCEPTION
by Irvin Rock

FROM QUARKS TO THE COSMOS
by Leon M. Lederman and David N. Schramm

HUMAN DIVERSITY
by Richard Lewontin

SLEEP
by J. Allan Hobson

THE SCIENCE OF WORDS
by George A. Miller

DRUGS AND THE BRAIN
by Solomon H. Snyder

If you would like to purchase additional volumes in the Scientific American Library, please send your order to:

Scientific American Library
41 Madison Avenue
New York, NY 10010